NUMERICAL COMPUTING: an introduction

LAWRENCE F. SHAMPINE
Sandia Laboratories and University
of New Mexico, Albuquerque, New Mexico

RICHARD C. ALLEN, JR.
Department of Mathematics and Statistics
University of New Mexico, Albuquerque, New Mexico

1973

W. B. SAUNDERS COMPANY
Philadelphia / London / Toronto

W. B. Saunders Company: West Washington Square
Philadelphia, PA 19105

12 Dyott Street
London, WC1A 1DB

833 Oxford Street
Toronto 18, Ontario

Numerical Computing ISBN 0-7216-8150-6

 Made in the United States of America. Press of W. B. Saunders Company. Library of Congress catalog card number 72-93122.

Print No: 9 8 7 6 5 4 3 2 1

We dedicate this book to

J. Todd and G. M. Wing

our teachers, colleagues, and friends.

We are happy to acknowledge assistance with the problems provided by M. K. Gordon and D. R. McLaughlin. We appreciate the care A. G. Conkle took with the typing of the manuscript and, with the aid of P. R. Chance, the seemingly interminable alterations.

CONTENTS

INTRODUCTION

This text examines the solution of some of the more common problems of numerical computation. It is intended for a semester course requiring only calculus and a modest acquaintance with FORTRAN programming as prerequisites. These constraints of background and time have important implications. The techniques developed are for the frequently occurring problem, not the pathologically difficult. By examining only a single, effective algorithm for each basic problem, the fundamental theory can be developed in a brief, elementary way. Along with the many exercises, solutions are provided which assist in developing mathematical and computational insight. Codes have been provided, implementing the various algorithms we shall present, in order to reduce the time otherwise required for programming and debugging.

We suggest that the chapters be read in the order: Floating Point Arithmetic, Systems of Linear Equations, Polynomial Interpolation, Integration, Roots of Nonlinear Equations, and Ordinary Differential Equations. This material fits nicely into a one semester course. For a two-quarter course or as time and special interests permit in a semester course, one could take up the remaining chapters, Least Squares Approximation and Cubic Splines. These last two chapters have been starred, as have some portions of II.3 and II.4. Non-starred material requires only elementary calculus; starred material also uses only calculus, but the arguments are more subtle.

The successful solution of a numerical problem has many facets. Among the more important are insight gleaned from the origin of the problem, guidelines from the theory of the algorithm chosen and computational experience with the algorithm, the computer being used, and the class of problems. There is little that we can say about insight except to emphasize that it is often the case that some feeling about the problem and the nature of the result anticipated may make the difference between a reasonable solution and utter nonsense. There is no substitute for common sense in evaluating and interpreting numerical results. Rarely does it happen that real problems are so simple that the theory describes them in detail. The theory is a guide to when an algorithm is applicable and to how effectively and accurately it can be expected to work. There are several reasons why computational experience is necessary. As we have indicated, the existing theory will not often carry us as far as realistic problems, though perhaps one can discuss pieces of a larger problem. This must then be supplemented by experience with similar problems which are simple enough that solutions are known. One must know what is realistic. Is it reasonable, for example, to try to integrate a differential equation to a relative accuracy of 10^{-10} on an IBM 360/67, a PDP-10, a CDC 6600? The answer depends on the particular machine and the general nature of the problem.

Many exercises are provided in varying degrees of difficulty; some are designed to get the reader to think about the text material and examples and to test his understanding, while others are purely computational in nature. Exercises are also used to some extent to introduce supplemental material. We suggest that the reader work as many of the exercises as possible. A "Solutions" part is provided, which is designed not only to provide answers to exercises but to give some mathematical and computational insight. The solutions should be read along with the text.

There is a great deal of difference between implementing formulas in FORTRAN and writing a production code, a fact not often appreciated by the beginning or casual programmer. For example, the code ZEROIN (a modification of an excellent code by Dekker) uses bisection and the secant rule to compute roots of nonlinear equations. A beginning programmer could implement the formulas easily, but the result would probably be far inferior. In ZEROIN careful attention has been given to problems of finite precision, overflow, convergence, and the like and the code has been carefully tested. For these reasons, as well as the previously mentioned limitation of time, we have chosen to provide codes. Within the constraints of being compatible on a number of machines and of being programmed in a simple, readable fashion, we believe these codes to be adequate for everyday use on non-pathological problems. Unless we state otherwise, all solutions and examples were computed using these codes on an IBM 360/67 using the WATFOR compiler.

This text is for the person who wants to do numerical computing. It is easy to arrive at a superficial understanding of an algorithm if the user confines himself to FORTRAN programming and to "canned" subroutines. We urge the reader in the exercises to study each algorithm through hand computation and to practice solving problems. It is essential that the reader understand how the codes provided work and precisely what they do. In the text, examples, and solutions we try to pass on some computational experience, but everyone must develop this for himself.

NOTATION AND SOME THEOREMS FROM THE CALCULUS

We assume that the reader is familiar with the topics normally covered in the undergraduate analytical geometry and calculus sequence. For purpose of reference, we present below some standard notation and theorems from the calculus.

NOTATION

$[a, b]$, the interval consisting of the real numbers x such that $a \leq x \leq b$.

(a, b), the interval consisting of real numbers x such that $a < x < b$.

$x \in [a, b]$, x lies in the interval $[a, b]$.

$$f'(x) = \frac{d}{dx} f(x).$$

$$f^{(n)}(x) = \frac{d^n}{dx^n} f(x).$$

$f \in C^1[a, b]$, f belongs to the class of functions having a first derivative continuous on the interval $[a, b]$.

$f \in C^n[a, b]$, f belongs to the class of functions having an nth derivative continuous on $[a, b]$.

$$g_x(x, y) = \frac{\partial}{\partial x} g(x, y), \quad g_y(x, y) = \frac{\partial}{\partial y} g(x, y), \quad g_{xy}(x, y) = \frac{\partial^2}{\partial x \, \partial y} g(x, y),$$

etc., denote partial differentiation.

$$\sum_{i=0}^{n} a_i = a_0 + a_1 + \cdots + a_n.$$

$$\prod_{i=0}^{n} a_i = a_0 \times a_1 \times \cdots \times a_n.$$

$.d_1 d_2 \cdots d_n(e)$ stands for $.d_1 d_2 \cdots d_n \times 10^e$, e.g., $.123(4)$ means $.123 \times 10^4$. This notation is used only in tables for readability.

$\doteq$, is approximately equal to.

$$\mathbf{v} = \begin{pmatrix} v_1 \\ v_2 \\ \cdot \\ \cdot \\ \cdot \\ v_n \end{pmatrix}, \text{ notation for vector.}$$

$q(h) = 0(h^k)$ as $h \to 0$, there are (unknown) constants h_0, K such that $|q(h)| \leq Kh^k$ for all $0 < h \leq h_0$.

THEOREMS

Theorem 1 Intermediate Value Theorem. Let $f(x)$ be a continuous function on the closed interval $[a, b]$. If for some number α and for some $x_1, x_2 \in [a, b]$ we have $f(x_1) \leq \alpha \leq f(x_2)$, then there is a point $c \in [a, b]$ such that

$$\alpha = f(c).$$

Theorem 2 Rolle's Theorem. Let $f(x)$ be continuous on the closed, finite interval $[a, b]$ and differentiable on (a, b). If $f(a) = f(b) = 0$, there is a point $c \in (a, b)$ such that

$$f'(c) = 0.$$

Theorem 3 Mean-Value Theorem for Integrals. Let $g(x)$ be a non-negative function integrable on the interval $[a, b]$. If $f(x)$ is continuous on $[a, b]$, then there is a point $c \in [a, b]$ such that

$$\int_a^b f(x)g(x)\,dx = f(c)\int_a^b g(x)\,dx.$$

Theorem 4 Mean-Value Theorem for Derivatives. Let $f(x)$ be continuous on the finite, closed interval $[a, b]$ and differentiable on (a, b). Then there is a point $c \in (a, b)$ such that

$$\frac{f(b) - f(a)}{b - a} = f'(c).$$

Theorem 5 Taylor's Theorem (with Remainder). Let $f(x)$ have a continuous derivative of order $n + 1$ on some interval (a, b) containing the points x and x_0. Set

$$f(x) = f(x_0) + \frac{f'(x_0)}{1!}(x - x_0) + \frac{f''(x_0)}{2!}(x - x_0)^2 + \cdots + \frac{f^{(n)}(x_0)}{n!}(x - x_0)^n + R_{n+1}(x).$$

Then there is a number c between x and x_0 such that

$$R_{n+1}(x) = \frac{f^{(n+1)}(c)}{(n+1)!}(x - x_0)^{n+1}.$$

Theorem 6 Let $f(x)$ be a continuous function on the finite, closed interval $[a, b]$. Then $f(x)$ assumes its maximum and minimum values on $[a, b]$; i.e., there are points $x_1, x_2 \in [a, b]$ such that

$$f(x_1) \leq f(x) \leq f(x_2)$$

for all $x \in [a, b]$.

Theorem 7 Integration by Parts. Let $f(x)$ and $g(x)$ be real valued functions with derivatives continuous on $[a, b]$. Then

$$\int_a^b f'(t)g(t)\,dt = f(t)g(t)\Big|_{t=a}^{t=b} - \int_a^b f(t)g'(t)\,dt.$$

Theorem 8 Fundamental Theorem of the Integral Calculus. Let $f(x)$ be continuous on the interval $[a, b]$, and let

$$F(x) = \int_a^x f(t)\,dt \quad \text{for all} \quad x \in [a, b].$$

Then $F(x)$ is differentiable on (a, b) and

$$F'(x) = f(x).$$

For discussions and proofs of these theorems, see references [5, 38, 41] at the end of this book.

Part I

FLOATING POINT ARITHMETIC

Computers use an approximation to the real number system called floating point arithmetic. We briefly present an idealized floating point arithmetic and study some of its consequences for practical computing. A very interesting and readable account of the interaction of the floating point number system with the solution of quadratic equations has been given by Forsythe [18]; this is a nice little paper to read in conjunction with the text. Henrici [23] gives another elementary treatment of floating point arithmetic which introduces the useful idea of a statistical treatment of errors. To pursue the subject in depth the reader ought to consult the text *Rounding Errors in Algebraic Processes* by J. H. Wilkinson [45]. Wilkinson's books are unmatched for their blend of theoretical advance, striking examples, practical insight, applications, and readability. We touch upon the subject of function evaluation subroutines. An elementary introduction to this interesting and important subject is found in Fike's book, *Computer Evaluation of Mathematical Functions* [16].

1. FLOATING POINT ARITHMETIC

BASIC CONCEPTS

Nearly all numerical computation on a digital computer is done in floating point arithmetic. This is a number system which uses a finite number of digits to approximate the real number system which we use in exact computation. Each non-zero floating point number y has the form

$$y = \pm .d_1 d_2 \cdots d_s \times 10^e.$$

At first we speak of decimal numbers because of their familiarity. There are a fixed number s of significant digits. There is a sign $+$ or $-$ and an exponent e which lies in a fixed range

$$m \leq e \leq M.$$

The number is normalized so that the first digit d_1 satisfies

$$1 \leq d_1 \leq 9.$$

Of course, the remaining digits satisfy

$$0 \leq d_k \leq 9 \qquad k = 2, 3, \ldots, s.$$

The number zero is in the floating point number system and we shall suppose it is represented by

$$+ .00 \cdots 0 \times 10^m$$

though practice varies. An example of a floating point decimal number system with $s = 1$, $m = -1$, $M = 1$ is the set of numbers

$$+ .1 \times 10^{-1}, + .2 \times 10^{-1}, \ldots, + .9 \times 10^{-1},$$

$$+ .1 \times 10^{0}, \;\; + .2 \times 10^{0}, \ldots, + .9 \times 10^{0},$$

$$+ .1 \times 10^{1}, \;\; + .2 \times 10^{1}, \ldots, + .9 \times 10^{1}$$

along with the negative of each of these numbers and $+ .0 \times 10^{-1}$ for zero. There are only 55 numbers in this floating point number system.

The portion $.d_1 d_2 \cdots d_s$ of the floating point number is called the fraction or mantissa or significand, and the number e is called the exponent.

We shall mention specific examples later, but most medium to large computers would have s about 9 and m and M of the order of -75 and $+75$, respectively. Since there are only finitely many floating point numbers to represent the real number system, each floating point number must represent many real numbers. Any real number $y \neq 0$ can be written as

$$y = \pm .d_1 d_2 \cdots d_s d_{s+1} \cdots \times 10^e$$

where $1 \leq d_1 \leq 9$ and in general infinitely many digits are required. (If there are two expansions, one ending in all 9's and the other 0's, use the one with 0's.) If $e > M$ or $e < m$, the number y is outside the range of the floating point system and cannot be represented at all. If it is inside the range, one way to represent y is by chopping off all digits after the first s to get

$$\text{fl}(y) = \pm .d_1 d_2 \cdots d_s \times 10^e.$$

Another is to add 5 to d_{s+1} and then chop; this is called rounding. Regardless of which procedure is used, we shall call this representation "rounded," though sometimes we shall want to distinguish the two.

What kind of error do we make in representing a real number in a floating point number system? The error is not uniform since the floating point numbers are not equally spaced. As our example shows, they are equally spaced between successive powers of 10. We shall see that this means the error is "small" in a relative sense. In general we mean by absolute error the quantity

$$\text{true value} - \text{approximate value}$$

and by relative error

$$\frac{\text{true value} - \text{approximate value}}{\text{true value}}.$$

Relative error is not defined if the true value is zero. Let us also define a unit roundoff in s digit decimal arithmetic to be

$$\begin{cases} u = \frac{1}{2}10^{1-s} & \text{rounded} \\ u = 10^{1-s} & \text{chopped} \end{cases}$$

with the appropriate choice depending on the mode of operation. Then we have the result:

Lemma. If y is a real number within the range of floating point numbers, then

$$\text{fl}(y) = y(1 + \delta) \quad \text{where} \quad |\delta| \leq u.$$

Another way to put this is

$$\left| \frac{y - \text{fl}(y)}{y} \right| = |\delta| \leq u$$

if $y \neq 0$.

PROOF. If

$$y = \pm .d_1 d_2 \cdots d_s d_{s+1} \cdots \times 10^e$$

and we chop to get

$$\text{fl}(y) = \pm .d_1 d_2 \cdots d_s \times 10^e,$$

then

$$\begin{aligned} y - \text{fl}(y) &= \pm .d_{s+1} d_{s+2} \ldots \times 10^{e-s} \\ &= \frac{\pm .d_{s+1} d_{s+2} \ldots \times 10^{e-s}}{\pm .d_1 d_2 \ldots \times 10^e} \times y \\ &= \delta y \end{aligned}$$

where

$$\delta = \frac{.d_{s+1} d_{s+2} \cdots}{.d_1 d_2 \ldots} \times 10^{-s}.$$

Now $1 \leq d_1 \leq 9$, and for all $k = s + 1, \ldots,$ we have $0 \leq d_k \leq 9$. This implies

$$.d_{s+1} d_{s+2} \cdots \leq 1,$$

$$.1 \leq .d_1 d_2 \cdots$$

so that

$$|\delta| \leq 10 \times 10^{-s} = u.$$

A similar argument gives the result for rounding instead of chopping.

Arithmetic in the floating point number system is to approximate that in the real number system. If y and z are in the floating point number system, they are also real numbers and we use $\oplus$, $\ominus$, $\otimes$, $\oslash$ to indicate the floating point approximations to the arithmetic operations $+, -, \times, /$. The product $y \times z$ is a real number but not in general a floating point number since y and z require s digits and their product could require $2s$ digits; for example, $.999 \times .999 = .998001$. About the best we could hope for is that a floating point arithmetic will produce the rounded result $\text{fl}(y \times z)$. So we assume an idealized arithmetic which for the basic arithmetic operations produce the rounded floating point numbers representing the exact results,

$$y \oplus z = \text{fl}(y + z),$$

$$y \ominus z = \text{fl}(y - z),$$

$$y \otimes z = \text{fl}(y \times z),$$

$$y \oslash z = \text{fl}(y/z),$$

providing they lie in the range of the floating point system. If the exponent e of the exact result is greater than M, we speak of overflow; if e is less than m, underflow. Computer systems monitor computations to see if the exponent range is exceeded and take appropriate action if it is. Usually one quits on overflow; frequently one sets the result to zero, gives a warning, and continues on underflow. This action on underflow may or may not be desirable, depending on the nature of the computations, so a warning is essential.

We could write the statements about floating point arithmetic as

$$y \circledcirc\!\!\text{op}\; z = (y \text{ op } z)(1 + \delta)$$

where $|\delta| \leq u$ and op $= +, -, \times, /$. This says that the floating point result is the exact result to within a unit roundoff error, and follows directly from the lemma. For real computers, this statement is still essentially true, although the value of u may need to be increased a little. Real addition can be less satisfactory, but we shall not go into the matter here.

Let us look at a simple example. Suppose we evaluate $(x - 1)^3$ in the form $((x - 3)x + 3)x - 1$ in 3-digit chopped decimal arithmetic with $-10 \leq e \leq 10$ at $x = 1.017$. The computations are in order:

i) $\text{fl}(1.017) = .101 \times 10^1 = r_0$

ii) $r_0 \ominus \text{fl}(3) = \text{fl}(.101 \times 10^1 - .300 \times 10^1) = -.199 \times 10^1 = r_1$

iii) $r_1 \otimes \text{fl}(x) = \text{fl}(-.199 \times 10^1 \times .101 \times 10^1) = \text{fl}(-.020099 \times 10^2)$
$= \text{fl}(-.20099 \times 10^1) = -.200 \times 10^1 = r_2$

iv) $r_2 \oplus \text{fl}(3) = \text{fl}(-.200 \times 10^1 + .300 \times 10^1) = .100 \times 10^1 = r_3$

v) $r_3 \otimes \text{fl}(x) = \text{fl}(.100 \times 10^1 \times .101 \times 10^1) = \text{fl}(.010100 \times 10^2)$
$= \text{fl}(.10100 \times 10^1) = .101 \times 10^1 = r_4$

vi) $r_4 \ominus \text{fl}(1) = \text{fl}(.101 \times 10^1 - .100 \times 10^1) = \text{fl}(.001 \times 10^1)$
$= \text{fl}(.100 \times 10^{-1}) = .100 \times 10^{-1}$.

Thus the computed value is $.100 \times 10^{-1}$. The true value is $(1.017 - 1)^3 = 4.913 \times 10^{-6}$. The relative error in the value of the polynomial is large:

$$\frac{4.913 \times 10^{-6} - 10^{-2}}{4.913 \times 10^{-6}} \doteq -2.034 \times 10^3.$$

An added complication to floating point arithmetic is that rather few computers use decimal arithmetic. A fraction

$$.d_1 d_2 \cdots d_s$$

means

$$d_1 \times 10^{-1} + d_2 \times 10^{-2} + \cdots + d_s \times 10^{-s}.$$

In the base β we require

$$1 \le d_1 \le \beta - 1$$
$$0 \le d_k \le \beta - 1 \qquad \text{for} \quad k = 2, \ldots, s$$

and the fraction means

$$d_1 \times \beta^{-1} + d_2 \times \beta^{-2} + \cdots + d_s \times \beta^{-s}.$$

Now we write

$$y = \pm .d_1 d_2 \cdots d_s \times \beta^e.$$

The most common values of β are 2, 8, 10, and 16. Some typical floating point arithmetics used in this text are

$\beta = 2, \quad s = 48, \quad m = -975, \quad M = 1071$ CDC 6600

$\beta = 2, \quad s = 28, \quad m = -128, \quad M = 127$ PDP-10

$\beta = 16, \quad s = 6, \quad m = -64, \quad M = 63$ IBM S/360

$\beta = 16, \quad s = 14, \quad m = -64, \quad M = 63$ IBM S/360

Base 16 arithmetic is called hexadecimal, and base 2 arithmetic is called binary. Notice that the IBM System 360 provides two sets of arithmetic. One system has more than twice as many digits as the other. The floating point system built into the computer is referred to as its single precision arithmetic. The FORTRAN language specifies that there will be another arithmetic of at least twice as many digits as single precision; it is usually called double precision and may be provided by either software or hardware. The IBM System 360 has its double precision built into the hardware. The CDC 6600 and the PDP-10 provide double precision arithmetic, too, but it is done by software. Typically, software double precision arithmetic takes considerably more than twice as long as hardware single precision arithmetic; hardware double precision takes considerably less than twice as long as single precision.

All the observations we have made about decimal arithmetic apply as well to other bases. Only the unit roundoff has to be redefined:

$$\begin{cases} u = \frac{1}{2}\beta^{1-s} & \text{rounded} \\ u = \beta^{1-s} & \text{chopped.} \end{cases}$$

The CDC 6600 uses chopped binary, the PDP-10 rounded binary, and the IBM S/360 chopped hexadecimal. Consideration of the unit roundoff and the error expressions for the basic operations shows that chopping instead of rounding is more serious for the larger bases like 10 and 16.

The product of two single precision numbers of s digits requires no more than $2s$ digits and so can be exactly represented in double precision. On many computers the full product is formed and then rounded to single precision as specified by our idealized arithmetic. A quantity of the form

$$\sum_{i=1}^{n} x_i y_i$$

is called an inner product. Errors in computing it come from two sources—the n multiplications and the $n - 1$ additions. If one multiplies the two single precision numbers x_i and y_i to form a double precision product and then does all the additions in double precision, he removes all the error due to multiplication. This is called accumulation of inner products. Most desk calculators do this automatically, and some computers do, too. A computer like the IBM 360/67 can do it but the FORTRAN does not, so one must use assembly language to accumulate inner products. This is not expensive because the only extra cost over single precision computation is in doing double precision additions; they are typically only a little more expensive than single precision additions (if the hardware is provided). In matrix computations inner products are a basic step and their accumulation can significantly improve overall accuracy.

IMPLICATIONS

Most numbers that we use in floating point computation must be presumed to be somewhat in error. They may be rounded from the values we have in mind by the input conversion process or they may be internally computed or manipulated with the resultant errors of inexact arithmetic. If we are interested in the value of a function $F(x)$, we must ask ourselves how this value is affected by errors in the argument x. Suppose x has a relative error of ϵ so that we actually use the value $x(1 + \epsilon)$. To gain some insight, suppose our routine for evaluating $F(x)$ is *exact* whatever the floating point argument. Thus we have the exact value of $F(x(1 + \epsilon))$ instead of $F(x)$. If F is differentiable, the error in the value of F caused by the error in x can be approximated by

$$F(x + \epsilon x) - F(x) \doteq \epsilon x F'(x).$$

(Theorem 4, p. 4). If we are interested in relative error, this leads to

$$\frac{F(x + \epsilon x) - F(x)}{F(x)} \doteq \epsilon x \frac{F'(x)}{F(x)}.$$

For example, suppose $F(x) = e^x$. Then the absolute error in the exponential due to an error in x is approximately

$$\epsilon x e^x$$

and the relative error is approximately

$$\epsilon x.$$

These errors can be serious if x is large. Thus, even if the routine EXP were perfect, there could be serious errors in using EXP(Y) for e^x when Y = fl(x) and x is a large real number. Another significant example is cosine x when x is near $\pi/2$. The absolute error is approximately

$$-\epsilon x \sin x \doteq -\epsilon \cdot \frac{\pi}{2} \cdot 1.$$

This is not troublesome, but the relative error is very much so; since $\cos \pi/2 = 0$,

$$-\epsilon x \frac{\sin x}{\cos x} \doteq -\epsilon \cdot \frac{\pi}{2} \cdot \frac{1}{0}.$$

Very small changes in x near $\pi/2$ cause very large relative changes in $\cos x$; we say the evaluation is unstable there. The accurate values

$$\cos 1.57079 = 0.63267949 \times 10^{-5}$$

$$\cos 1.57078 = 1.63267949 \times 10^{-5}$$

show how a very small change in the argument can have a profound effect on the function value. Obviously, the problem discussed in II.3 of finding a simple root of $F(x) = 0$, that is, a number α such that $F(\alpha) = 0$ and $F'(\alpha) \neq 0$, is going to be sensitive to machine arithmetic in just this way. The point of the analysis and examples is that some problems are extremely sensitive to their data, and so the use of rounded arguments can have a serious effect even if we presume that the remaining computations are exact.

Even the idealized arithmetic we have postulated behaves quite differently from the real number system. In some respects they agree, e.g., the commutative laws

$$x \otimes y = y \otimes x,$$

$$x \oplus y = y \oplus x$$

are valid. Inequalities like

$$\text{if} \quad u \leq v, \qquad \text{then} \qquad u \oplus w \leq v \oplus w$$

hold because $u + w \leq v + w$ and the floating result is the rounded version of this inequality. Some laws are almost true. In an exercise, we ask the reader to produce x, y, z such that

$$(x \otimes y) \otimes z \neq x \otimes (y \otimes z),$$

that is, the associative law of multiplication $(xy)z = x(yz)$ does not hold. We remark that it is not even necessary that both quantities exist, e.g., $x \otimes y$ may underflow or overflow while $x \otimes (y \otimes z)$ does not. Still, we know that if no quantity exceeds the exponent range,

$$x \otimes y = xy(1 + \delta_1),$$
$$(x \otimes y) \otimes z = xyz(1 + \delta_1)(1 + \delta_2)$$

and similarly

$$x \otimes (y \otimes z) = xyz(1 + \delta_3)(1 + \delta_4).$$

Then

$$\frac{x \otimes (y \otimes z)}{(x \otimes y) \otimes z} = \frac{(1 + \delta_3)(1 + \delta_4)}{(1 + \delta_1)(1 + \delta_2)} = 1 + \epsilon,$$

where ϵ is "small" and can, of course, be explicitly bounded. Hence, multiplication is "nearly" associative. Addition is quite another matter. It can easily happen that $(x \oplus y) \oplus z$ is very different from $x \oplus (y \oplus z)$. It is also the case that the distributive law $x(y + z) = xy + xz$ connecting multiplication and addition is not valid in floating point arithmetic.

Addition is the trouble spot in floating point arithmetic as far as relative error is concerned. Our idealized arithmetic has the sum of two numbers obtained with a low relative error, but this is not true of three numbers. Suppose we want to compute

$$x_0 + x_1 + x_2 + \cdots + x_n.$$

Because addition is not associative, we have to specify exactly how this is to be done. For example, let

$$s_0 = x_0$$
$$s_k = s_{k-1} + x_k \qquad k = 1, \ldots, n.$$

Then s_n is the desired sum. Let s_k^* denote the computed values:

$$s_0^* = x_0 = s_0$$
$$s_k^* = s_{k-1}^* \oplus x_k \qquad k = 1, \ldots, n.$$

In our idealized arithmetic

$$s_1^* = x_0 \oplus x_1 = x_0(1 + \delta_1) + x_1(1 + \delta_1)$$

and

$$s_2^* = s_1^* \oplus x_2 = s_1^*(1 + \delta_2) + x_2(1 + \delta_2)$$
$$= x_0(1 + \delta_1)(1 + \delta_2) + x_1(1 + \delta_1)(1 + \delta_2) + x_2(1 + \delta_2).$$

The error

$$s_2^* - s_2 =$$
$$x_0[(1+\delta_1)(1+\delta_2)-1]+x_1[(1+\delta_1)(1+\delta_2)-1]+x_2[(1+\delta_2)-1]$$

need not be small compared to s_2, for all we know is that each $|\delta_i| \leq u$. What if $s_2 = 0$ but $s_2^* \neq 0$? Then there is no bound on the relative error even when summing just three numbers. If we continue this analysis, we see that

$$\begin{aligned}
s_n^* - s_n = {} & x_0[(1+\delta_1)(1+\delta_2)\ldots(1+\delta_n)-1] \\
& + x_1[(1+\delta_1)(1+\delta_2)\ldots(1+\delta_n)-1] \\
& + x_2[(1+\delta_2)(1+\delta_3)\ldots(1+\delta_n)-1] \\
& \quad \vdots \\
& + x_n[(1+\delta_n)-1].
\end{aligned}$$

In general there is no useful bound on the relative error because of the possibility that the final sum is small. This expression suggests that as a rule of thumb one ought to add the numbers in order of increasing magnitude, $|x_0| \leq |x_1| \leq \cdots \leq |x_n|$, because the bounds on the error factors are smaller for the terms added later. If all the terms are of the same sign, the partial sums s_k grow in magnitude and we would expect the relative error to be reasonable. Suppose all the x_i above have the same sign so that

$$\begin{aligned}
|s_n^* - s_n| \leq {} & |x_0|\,[(1+u)^n - 1] \\
& + |x_1|\,[(1+u)^n - 1] \\
& + |x_2|\,[(1+u)^{n-1} - 1] \\
& + \ldots + |x_n|\,[(1+u) - 1] \\
\leq {} & \sum_{i=0}^{n} |x_i| \cdot [(1+u)^n - 1] \\
\leq {} & \left| \sum_{i=0}^{n} x_i \right| \cdot [(1+u)^n - 1].
\end{aligned}$$

Then

$$\left| \frac{s_n^* - s_n}{s_n} \right| \leq (1+u)^n - 1.$$

We would like a convenient bound for the expression $(1+u)^n - 1$. If we restrict n, we obtain the following result.

Lemma. If $0 \leq nu \leq b < 1$, then

$$(1 + u)^n \leq 1 + \frac{nu}{1 - b}.$$

PROOF. The result is proved by induction on n. The case $n = 1$,

$$1 + u \leq 1 + \frac{u}{1 - b},$$

is obviously true since $0 \leq b < 1$. Suppose the inequality is true for $n = k$. Then multiplying by $1 + u$, we have

$$\begin{aligned}(1 + u)(1 + u)^k &\leq (1 + u)\left(1 + \frac{ku}{1 - b}\right)\\ &= 1 + \frac{ku}{1 - b} + u + \frac{ku^2}{1 - b}\\ &\leq 1 + \frac{ku}{1 - b} + u\left(1 + \frac{b}{1 - b}\right)\\ &= 1 + \frac{(k + 1)u}{1 - b}\end{aligned}$$

which establishes the result for $n = k + 1$.

If we apply this to the bound above, we see that if $nu \leq b < 1$, then

$$\left|\frac{s_n - s_n^*}{s_n}\right| \leq \frac{nu}{1 - b}.$$

If there are not too many terms and if they are all of the same sign, then summation is accurate in the relative error sense. Unfortunately, there are many important computations which involve a great many additions, enough to cause trouble even if all terms are of the same sign. Before looking at some examples, let us look more closely at two extreme cases, the addition of two numbers of very different size and the subtraction of two nearly equal numbers.

The equation $z + y = z$ has only the solution $y = 0$ in the real number system, but it has many solutions in the floating point number system. Suppose $z = .d_1d_2 \cdots d_s \times \beta^e$ and $y = .f_1f_2 \cdots f_s \times \beta^p$. The floating point result is the rounded version of the exact result. To add these numbers we must adjust the exponents to be the same, that is, align the "decimal" point. Suppose $p + s = e$; then

$$\begin{aligned} z &= .d_1d_2 \cdots d_s00 \cdots \times \beta^e\\ y &= .00 \quad \cdots 0f_1f_2 \cdots \times \beta^e\\ \hline z + y &= .d_1d_2 \cdots d_sf_1f_2 \cdots \times \beta^e\end{aligned}$$

If we add z to y and chop to s figures, we see that the result is z. Indeed, this is true for any p with $p + s \leq e$. A similar result holds for rounding. The point is that if z is sufficiently large compared to y, the digits of y do not participate in the addition. The floating point sum is quite reasonable since it is accurate in a relative error sense.

Another aspect of addition is the phenomenon of cancellation. Suppose we subtract two numbers z and y which have the same exponent and agree in their first k leading figures:

$$\begin{array}{rl} z = & .d_1 d_2 \cdots d_k d_{k+1} \cdots d_s \times \beta^e \\ y = & .d_1 d_2 \cdots d_k f_{k+1} \cdots f_s \times \beta^e \\ \hline z - y = & .00 \quad \cdots 0 r_{k+1} \quad \cdots r_s \times \beta^e \end{array}$$

The floating point result is the exact result after it has been normalized and rounded:

$$\text{fl}(z - y) = .r_{k+1} \cdots r_s 0 \cdots 0 \times \beta^{e-k}.$$

The leading digits have cancelled out and the result is exact. When cancellation takes place the computation is exact, so it does not itself cause numerical difficulties, but the answer is smaller and frequently errors that were already present become more serious in a relative error sense.

An important part of classical applied mathematics is deriving solutions to problems in terms of series. When one is working to high accuracies or if the series converges slowly, it is necessary to add up a great many terms to sum the series. Series are less important to modern applied mathematics, but they are easy to visualize and, as we shall indicate, they illustrate what can happen in several processes we shall examine later in this text. If one must add sufficiently many terms, the accumulation of the individually small relative errors can be serious. For example, if we sum

$$1, \underbrace{\tfrac{1}{2}u, \tfrac{1}{2}u, \ldots, \tfrac{1}{2}u}_{k \text{ terms}}$$

from left to right, we see that $1 \oplus \frac{1}{2}u = 1$ at every step. The final sum is 1 and the absolute error is $\frac{ku}{2}$. When k is large, this is a serious error. If there is cancellation, the accumulated error is generally large in a relative sense. An extreme case would be the summation of

$$1, \tfrac{1}{2}u, \ldots, \tfrac{1}{2}u, -1$$

for which the computed sum is zero. A classic example is the divergent series

$$\sum_{n=1}^{\infty} \frac{1}{n}.$$

The point is that in floating point arithmetic this series has a finite sum. The terms $\dfrac{1}{k+1}$ eventually become so small compared to the partial sum

$$s_k = \sum_{n=1}^{k} \frac{1}{n}$$

that $s_{k+1}^* = s_k^* \oplus \dfrac{1}{k+1} = s_k^*$. To be fair, we should point out that the partial sums grow very slowly. An accurate value for the sum of the first 10^6 terms is 14.39273 . . . . Generally, if one is to obtain an accurate value of $\sum_{n=0}^{\infty} a_n$, the series must converge fairly rapidly even if all terms are of the same sign. The trick of summing in the reverse direction can be very helpful, since convergent series have terms which tend to decrease steadily.

A classic result says that if $a_n \geq 0$ and $a_n \geq a_{n+1}$ for all n, then the alternating series

$$s = \sum_{n=0}^{\infty} (-1)^n a_n$$

converges and the error of a partial sum

$$s_N = \sum_{n=0}^{N} (-1)^n a_n$$

is of the same sign as the next term, namely $(-1)^{N+1}$, and

$$|s - s_N| \leq a_{N+1}.$$

One is inclined to think of such a series as ideal, but it can be very troublesome numerically. Suppose we have to compute the terms a_n and we do this with error, thus arriving at $a_n + \epsilon_n$. If the a_n decrease slowly so that $a_n \doteq a_{n+1}$, we see that

$$s_{2m+1} = (a_0 - a_1) + (a_2 - a_3) + \cdots + (a_{2m} - a_{2m+1}) \doteq 0.$$

Then

$$s_{2m+2} = s_{2m+1} + a_{2m+2} \doteq a_{2m+2}$$

and the computed value

$$s_{2m+2}^* \doteq a_{2m+2} + \epsilon_{2m+2}.$$

However, there is severe cancellation at the next step:

$$s_{2m+3} = s_{2m+2} - a_{2m+3} \doteq 0$$

and

$$s^*_{2m+3} \doteq 0 + \epsilon_{2m+2} - \epsilon_{2m+3}.$$

The errors in the computed values of the terms become very prominent.

The approximation of

$$I(f) = \int_a^b f(x)\,dx$$

leads to problems just like those associated with series. In II.2 we discuss more potent methods, but the flavor is conveyed by approximating $I(f)$ by a Riemann sum

$$I(f) \doteq \sum_{j=1}^{n} hf(a + jh)$$

where $h = \dfrac{b-a}{n}$. The solution of the differential equation

$$y'(x) = f(x, y(x)),$$
$$y(a) = A$$

also leads to these difficulties. Although we discuss more potent methods in II.4, Euler's method illustrates the matter. One chooses an h and makes up a table of approximate values $y_j \doteq y(a + jh)$ where

$$y_0 = A,$$
$$y_{j+1} = y_j + hf(a + jh, y_j).$$

In each of these processes, one has many additions of quantities with errors in them and the possibility of cancellation making the errors very significant in a relative sense.

In III we discuss the numerical solution of a system of linear equations of which the simplest case is

$$ax = b.$$

We compute an approximate solution z and measure how well it satisfies the equation by computing the residual

$$r = b - az.$$

If a and b are floating point numbers,

$$\begin{aligned} a \otimes z &= az(1 + \delta_1), \\ b \ominus (a \otimes z) &= [b - az(1 + \delta_1)](1 + \delta_2) \\ &= [r - \delta_1 az](1 + \delta_2) \\ &= r + \delta_2 r - \delta_1 az - \delta_1\delta_2 az \\ &= r + \delta_2 r - \delta_1 b - \delta_1\delta_2 b. \end{aligned}$$

The point here is that the term $\delta_1 b$ might be as large as ub, which may well be as large as r itself. We must turn to higher precision to compute a value of r which will meaningfully measure the accuracy of z.

EXERCISES

1. Prove that in our idealized floating point arithmetic

$$x \otimes y = y \otimes x,$$
$$x \oplus y = y \oplus x.$$

2. In the chopped decimal floating point arithmetic with numbers $\pm .d_1 d_2 d_3 \times 10^e$ with $-100 \le e \le 100$, construct examples to show that in general:
   i) $(x \otimes y) \otimes z \neq x \otimes (y \otimes z)$
   ii) $(x \oplus y) \oplus z \neq x \oplus (y \oplus z)$
   iii) $x \otimes (y \oplus z) \neq (x \otimes y) \oplus (x \otimes z)$
   iv) $(x \oplus y) \oplus z$ can have a "large" relative error.

3. Work out a bound for the relative error ϵ in the associative law for multiplication:

$$(x \otimes y) \otimes z = x \otimes (y \otimes z)(1 + \epsilon)$$

4. Suppose that $z = .180 \times 10^2$ is an approximate solution of $ax = b$ for $a = .111 \times 10^0$, $b = .200 \times 10^1$. Using the arithmetic of problem 2, compute the residual. Compute the residual in double precision and in exact arithmetic. Compare all three results.

5. Suppose we wish to evaluate $\sqrt{x}$ but the machine representation of x is $\text{fl}(x) = x(1 + \epsilon)$ where the relative error ϵ is small. Show that if the routine SQRT is exact, then $\sqrt{x}$ is obtained with a small relative error. If the routine is such that for any floating point number y, $\text{SQRT}(y) = \sqrt{y}(1 + \delta)$ where δ is small, show that $\sqrt{x}$ is still obtained with a small relative error.

6. Using the arithmetic of problem 2, find floating point numbers u, v, and w for which exponent overflow occurs during the calculation of $u \otimes (v \otimes w)$ but not during the calculation of $(u \otimes v) \otimes w$.

7. Is $u \oslash v = u \otimes (1 \oslash v)$ an identity for all floating point numbers u and $v \neq 0$ such that no exponent overflow or underflow occurs?

8. Compute an approximation to e^x via the sequence of partial sums

$$s_0, s_1, s_2, \ldots$$

where

$$s_k = \sum_{j=0}^{k} \frac{x^j}{j!}.$$

Note that $s_{k+1} = s_k + P_{k+1}$ where $P_{k+1} = xP_k/(k+1)$ and $s_0 = 1$, $P_0 = 1$. For $x = -10$, find the smallest value of n such that $s_n = s_{n+1} = s_{n+2} = \cdots$. Compute the absolute and relative errors in your approximation. In computing the absolute error, evaluate e^x in double precision and form the difference $|s_n - e^x|$ in double precision. Does this look like a good way to calculate e^x?

Part II

NUMERICAL ANALYSIS

This part is devoted to the numerical solution of problems in analysis, that is, problems in which the concept of limit plays a central role. Such problems are especially difficult because the floating point number system is finite and one cannot directly imitate the limits in the real number system used by traditional mathematical solutions. As a result, basic solution techniques of analysis must be reformulated or approximated in a computationally feasible manner. Historically, this task has been so important as to cause this whole area of applied mathematics to be called numerical analysis. There are other important aspects to numerical analysis (see for example the next part of this text), so we have preferred the term *numerical computing* for the title of this text.

Chapter II.1 is devoted to interpolation and approximation. The first portion, II.1.1, is concerned with polynomial interpolation, which is the subject of a vast literature. We use divided differences since they seem best for machine computation [28], but there is an algorithm due to Aitken which is very convenient for hand computation. More advanced texts like Henrici [23] and Isaacson and Keller [26] discuss this algorithm as well as other parts of the theory of polynomial interpolation. The treatment of least squares approximation in II.1.2 is restricted to polynomials in a single real variable, for which there is a special procedure. The interesting survey by Davis [12] will lead the reader into some of the theory and application of least squares methods. We make further reference to the general problem in Part III. There is a rapidly growing literature on splines, the subject of II.1.3. The text of Ahlberg, Nilson, and Walsh [2] is a good general reference.

The adaptive quadrature scheme developed in Chapter II.2 for computing integrals over a finite interval seems to be one of the most effective approaches for integrands which are continuous. Our treatment is based on codes and studies by McKeeman et al. [31, 32] and Lyness [30]. Another method due to Romberg is also popular, but does not seem to perform as well in comparative tests [8, 27]. An exposition of Romberg's method can be found in *Numerical Integration* by Davis and Rabinowitz [13]. This text also treats Gaussian quadratures, which are frequently the most effective way to handle integrands which become infinite or intervals which are infinite. A very comprehensive collection of formulas can be found in *Gaussian Quadrature Formulas* by Stroud and Secrest [40].

Deeper studies of the solutions of nonlinear equations treated in Chapter II.3 can be found in references [3, 24, 36, 37, 42, 46]. The references [14, 43, 45] are particularly suitable for following this text.

Solving ordinary differential equations and systems of linear equations are the two most important problems treated in this text. It strikes the eye that Chapter II.4 is twice the length of the corresponding chapter on linear equations. Why is this so? The theory of linear equations, to the extent we require, can be quickly developed in an entirely elementary way. Many of the elements are probably already familiar to the reader. The basic theory of ordinary differential equations, however, is much more complex and many of our readers may be unfamiliar with it. Because of this we must develop or state quite a bit of elementary background material. For example, we describe what differential equations are and give some basic results about solutions, such as when they exist and, if they exist, when there is only one. An important practical matter is to develop facility in writing differential equations in a standard form suitable for our code, and we must spend some time discussing this matter.

The theory of the numerical solution of ordinary differential equations is less developed than the theory of solving linear equations. This is partly because we approximate continuous functions instead of numbers and partly because there is a wider variety of behavior of solutions possible. At this time, there is a substantial element of art to the numerical solution of differential equations. The consensus of workers in the field seems to be that the most effective method for typical non-stiff problems is a variable step, variable order Adams code, with a possible competitor being rational extrapolation of a modified midpoint rule. Working computer codes for these methods are quite complex, and a development of the methods sufficiently detailed to enable the reader to understand the codes would itself require a text larger than the present one. The advanced text by Gear [22] develops and gives a good Adams code and discusses extrapolation methods. A particularly well documented extrapolation code is found in Fox's article [20]. Historically, much attention has been given to Runge-Kutta methods. They seem to be competitive with the preceding methods if one does not ask for much in the way of accuracy [25]. This is appropriate to our text, but the compelling reason for our developing Runge-Kutta methods is that the theory is simple enough so that we can give a reasonably complete development which is still elementary and not too long. Given that one is going to develop a Runge-Kutta code, we have presented a very complete treatment and a highly effective variant of the Runge-Kutta family of codes.

Elementary texts frequently develop a fourth order Adams code and compare it to a fourth order Runge-Kutta code. Usually the Runge-Kutta approach is dismissed with the statement that one step of the Adams code requires half the work of one step of the Runge-Kutta code, and so is obviously more efficient. This is silly because it does not consider the possibility, which is a fact, that the Runge-Kutta code can operate with a step twice as large. The advanced text by Ceschino and Kuntzmann [9] makes a fair comparison; neither is to be preferred on grounds of computational effort. The Runge-Kutta code is universally conceded to be preferred on grounds of convenience. The virtue of the Adams method appears when one goes to higher orders and especially to variable order codes.

1. INTERPOLATION AND APPROXIMATION

1.1. POLYNOMIAL INTERPOLATION

Basic Concepts

In this chapter we treat the important problem of approximating one function $f(x)$ by another, "more appropriate," function $g(x)$. There are several reasons why one might want to do this. For example, it might be desirable to replace a function $f(x)$ which is difficult to evaluate or manipulate, like $\log x$, $\sin x$, or $\operatorname{erf} x$, by a function $g(x)$ which is simpler and more easily handled, such as a polynomial. This idea is the foundation of methods for numerical integration and differentiation, finding roots of nonlinear equations, solving ordinary differential equations, indeed, for most of classical numerical analysis. Assuming that $f(x)$ is known for a discrete set of arguments or "nodes," say $x_0, x_1, \ldots, x_n$, our immediate task in this section is to construct a function which approximates f at other x values. There are many types of approximating functions $g(x)$, and which one to use depends to a large extent on the character of the data and on how one intends to use the approximation. The most common approximating functions are polynomials, and the interpolating polynomial is probably the most easily constructed and the most widely used.

First of all, what is an interpolating polynomial for the function $f(x)$? Suppose we have values for the real function $f(x)$ at the distinct nodes $x_0, x_1, \ldots, x_n$. For convenience we use the notation $f(x_k) = f_k$. A polynomial $P(x)$ is said to interpolate $f(x)$ on the nodes $x_0, x_1, \ldots, x_n$ if

$$P(x_k) = f_k, \qquad k = 0, 1, \ldots, n. \tag{1}$$

Of course, it is our hope that if $P(x)$ agrees with $f(x)$ at these nodes, then it is close to $f(x)$ in value elsewhere. It is not clear what degree is appropriate for $P(x)$. In general, it is not possible to satisfy all $(n + 1)$ conditions of (1) unless the degree of $P(x)$ is at least n. On the other hand, there are many polynomials which satisfy (1) if we allow the degree to be greater than n. There is, however, one and only one polynomial of degree not exceeding n which satisfies (1). We shall now prove this last statement, first establishing the existence of such a polynomial and then showing that it is unique. The proofs of the other statements are left to the exercises.

Consider a polynomial of degree n written in the form

$$P_n(x) = a_0 + (x - x_0)a_1 + (x - x_1)(x - x_0)a_2 + \cdots + (x - x_{n-1})(x - x_{n-2}) \cdots (x - x_0)a_n$$

where the x_i are the nodes of (1). As we successively require $P_n(x)$ to satisfy the interpolating conditions (1), the coefficients a_i are determined. From the first condition, $f_0 = P_n(x_0)$, we have

$$f_0 = P_n(x_0) = a_0.$$

The second condition, $f_1 = P_n(x_1)$, requires that

$$f_1 = P_n(x_1) = a_0 + (x_1 - x_0)a_1$$

from which we find

$$a_1 = \frac{f_1 - a_0}{x_1 - x_0} = \frac{f_1 - f_0}{x_1 - x_0}.$$

This is well defined since $x_1 \neq x_0$. In general, suppose we have determined $a_0, a_1, \ldots, a_i$ and satisfied

$$P_n(x_k) = f_k, \qquad k = 0, 1, \ldots, i.$$

Then

$$P_n(x_{i+1}) = f_{i+1} = a_0 + (x_{i+1} - x_0)a_1 + (x_{i+1} - x_1)(x_{i+1} - x_0)a_2 + \cdots + (x_{i+1} - x_i)(x_{i+1} - x_{i-1}) \cdots (x_{i+1} - x_0)a_{i+1}$$

from which we can solve for a_{i+1} since the factor multiplying a_{i+1} is non-zero. This process is continued through a_n, at which time we have constructed a polynomial of degree at most n (any of the a_i, in particular a_n, could be zero) which satisfies (1). In what follows we shall refer to $P_n(x)$ as a polynomial of degree n.

Example. Given the function $f(x) = \sin(x)$ and the nodes $x = 0$, $\pi/2$, π, let us construct $P_2(x)$. Since

$$P_2(x) = a_0 + (x - 0)a_1 + (x - 0)\left(x - \frac{\pi}{2}\right)a_2,$$

we have

$$P_2(0) = \sin(0) = 0 = a_0 + 0 \times a_1 + 0 \times a_2,$$

hence

$$a_0 = 0.$$

Also

$$P_2\left(\frac{\pi}{2}\right) = \sin\left(\frac{\pi}{2}\right) = 1 = 0 + \left(\frac{\pi}{2} - 0\right) a_1 + 0 \times a_2,$$

$$P_2(\pi) = \sin(\pi) = 0 = 0 + (\pi - 0)\ a_1\ + (\pi - 0)\left(\pi - \frac{\pi}{2}\right) a_2,$$

imply

$$a_1 = \frac{2}{\pi}, \quad a_2 = -\frac{4}{\pi^2}.$$

So

$$P_2(x) = 0 + \frac{2}{\pi}x - \frac{4}{\pi^2}x\left(x - \frac{\pi}{2}\right).$$

Note that $P_2(x)$ can be written in the perhaps more familiar, but sometimes less useful, form

$$P_2(x) = -\frac{4}{\pi^2}x^2 + \frac{4}{\pi}x.$$

An important observation is that the polynomial of degree k,

$$P_k(x) = a_0 + (x - x_0)a_1 + (x - x_1)(x - x_0)a_2 + \cdots + (x - x_{k-1}) \cdots (x - x_0)a_k, \quad (2)$$

is a polynomial which interpolates on the nodes $x_0, x_1, \ldots, x_k$. So, when we construct $P_n(x)$, we construct $P_k(x)$ at the same time for each $k = 0, 1, \ldots, n - 1$. This observation will prove valuable when we do not know in advance how many data points we shall require to adequately approximate $f(x)$. In the previous example $P_0(x) = 0$ interpolates $f(x) = \sin(x)$ on the node $x_0 = 0$ and $P_1(x) = 0 + \frac{2}{\pi}x$ interpolates on the nodes $x_0 = 0$, $x_1 = \pi/2$.

To show that $P_n(x)$ is unique, suppose $Q(x)$ is any other polynomial of degree at most n which interpolates $f(x)$ on the nodes $x_0, x_1, \ldots, x_n$, i.e., satisfies the conditions

$$Q(x_k) = f_k \qquad k = 0, 1, \ldots, n.$$

Then the difference $D(x) = P_n(x) - Q(x)$ is a polynomial of degree at most n which we can write as

$$D(x) = \gamma x^r + \text{lower degree terms}, \qquad \gamma \neq 0,$$

where $r \leq n$. This polynomial has $n + 1$ distinct zeros since

$$D(x_k) = P_n(x_k) - Q(x_k) = f_k - f_k = 0, \qquad k = 0, 1, \ldots, n.$$

If the reader is familiar with the theorem that a polynomial of degree r has only r zeros, we have an immediate contradiction implying $D(x) \equiv Q(x)$. Another proof uses Rolle's theorem, which states that between successive zeros of $D(x)$, the derivative $D'(x)$ must vanish at least once. This implies that

$$D'(x) = r\gamma x^{r-1} + \text{lower degree terms}$$

has at least n distinct zeros. If we repeat this argument several times, we finally arrive at the statement that the constant function

$$D^{(r)}(x) \equiv r!\,\gamma \ (\neq 0)$$

has at least $n + 1 - r > 0$ distinct zeros, which is impossible. The contradiction implies that $P_n(x)$ is the *only* polynomial of degree n which satisfies (1). As we shall see, there are many forms of the interpolating polynomial of degree n satisfying (1). The above remarks, however, assure us that they are just different representations of the same polynomial.

The coefficients of $P_n(x)$, the a_i, are called divided differences. It is clear from their construction that each a_i depends only on the function values $f_0, f_1, \ldots, f_i$. This is reflected in the common notations

$$a_i \equiv f[x_0, x_1, \ldots, x_i] = f_{0,1,\ldots,i}$$

which we shall henceforth use. In order to analyze the properties of the interpolating polynomial and to devise effective algorithms for computation, we shall establish two facts concerning divided differences. First, the *value* of a_i does not depend on the order in which the nodes are used, although the *construction* above does. This is exemplified in the case of a_1 for

$$f[x_0, x_1] = \frac{f(x_1) - f(x_0)}{x_1 - x_0} = \frac{f(x_0) - f(x_1)}{x_0 - x_1} = f[x_1, x_0].$$

Second, divided differences satisfy the simple relation

$$f[x_0, x_1, \ldots, x_i] = \frac{f[x_0, x_1, \ldots, x_{i-1}] - f[x_1, x_2, \ldots, x_i]}{x_0 - x_i}. \tag{3}$$

Again, the definition of a_1 furnishes a trivial example:

$$a_1 = f[x_0, x_1] = \frac{f[x_0] - f[x_1]}{x_0 - x_1} = \frac{f(x_0) - f(x_1)}{x_0 - x_1}.$$

It is not difficult to establish these facts, although the notation is perhaps forbidding. The proofs are rather interesting since they use only the fact that just one polynomial of degree n interpolates on $n + 1$ distinct nodes, and that two apparently different polynomials satisfying this condition are in fact the same polynomial written in different ways. At a first reading the student might prefer to skip over the proofs. As we saw above, the polynomial $P_n(x)$ arising from interpolation on the set $\{x_0, x_1, \ldots, x_n\}$ in the order $x_0, x_1, \ldots, x_n$ is

$$P_n(x) = f[x_0] + (x - x_0)f[x_0, x_1] + (x - x_1)(x - x_0)f[x_0, x_1, x_2] + \cdots + (x - x_{n-1}) \cdots (x - x_0)f[x_0, x_1, \ldots, x_n]. \quad (4)$$

In the same way, if we interpolate on the set in a different order $x_{i_0}, x_{i_1}, \ldots, x_{i_n}$, we find that

$$P_n(x) = f[x_{i_0}] + (x - x_{i_0})f[x_{i_0}, x_{i_1}] + (x - x_{i_1})(x - x_{i_0})f[x_{i_0}, x_{i_1}, x_{i_2}] + \cdots + (x - x_{i_{n-1}}) \cdots (x - x_{i_0})f[x_{i_0}, x_{i_1}, \ldots, x_{i_n}]. \quad (5)$$

These two polynomials both satisfy (1), are of degree n, and hence must be the same polynomial written in two different ways. If we notice from (4) that

$$P_n(x) = x^n f[x_0, x_1, \ldots, x_n] + \text{lower degree terms}$$

and from (5) that

$$P_n(x) = x^n f[x_{i_0}, x_{i_1}, \ldots, x_{i_n}] + \text{lower degree terms},$$

we find

$$f[x_0, x_1, \ldots, x_n] = f[x_{i_0}, x_{i_1}, \ldots, x_{i_n}]$$

because the two forms are identical. So the order of the arguments in a divided difference is immaterial and does not affect its value.

To establish (3), we write down the two forms of the interpolating polynomial arising from interpolation in the orders $x_0, x_1, \ldots, x_n$ and $x_1, x_2, \ldots, x_n, x_0$:

$$\begin{aligned} P_n(x) &= f[x_0] + (x - x_0)f[x_0, x_1] + \cdots \\ &\quad + (x - x_{n-2}) \cdots (x - x_0)f[x_0, x_1, \ldots, x_{n-1}] \\ &\quad + (x - x_{n-1}) \cdots (x - x_0)f[x_0, x_1, \ldots, x_n] \\ &= f[x_1] + (x - x_1)f[x_1, x_2] + \cdots \\ &\quad + (x - x_{n-1}) \cdots (x - x_1)f[x_1, x_2, \ldots, x_n] \\ &\quad + (x - x_n) \cdots (x - x_1)f[x_1, x_2, \ldots, x_n, x_0]. \end{aligned}$$

Multiply the first form by $(x - x_n)$, the second by $(x - x_0)$, and subtract to arrive at

$$(x - x_n)P_n(x) - (x - x_0)P_n(x) = \{(x - x_n)f[x_0] - (x - x_0)f[x_1]\} + \cdots$$
$$+ \{(x - x_n)(x - x_{n-2}) \cdots (x - x_0)f[x_0, x_1, \ldots, x_{n-1}]$$
$$- (x - x_0)(x - x_{n-1}) \cdots (x - x_1)f[x_1, x_2, \ldots, x_n]\}$$
$$+ \{(x - x_n)(x - x_{n-1}) \cdots (x - x_0)f[x_0, x_1, \ldots, x_n]$$
$$- (x - x_n) \cdots (x - x_1)(x - x_0)f[x_1, x_2, \ldots, x_n, x_0]\}.$$

Because of the previous result that the order of arguments does not affect the value of a divided difference, the last term in braces vanishes. Then

$$(x - x_n)P_n(x) - (x - x_0)P_n(x) = (x_0 - x_n)P_n(x)$$
$$= x^n(f[x_0, x_1, \ldots, x_{n-1}] - f[x_1, x_2, \ldots, x_n])$$
$$+ \text{lower degree terms.}$$

Dividing by $(x_0 - x_n)$ yields still another form of $P_n(x)$. Once again, equating the coefficients of x^n establishes the result we want.

Let us now develop some error expressions; in particular, we develop a practical method for estimating the error $f(x) - P_n(x)$. If we think of x as being a fixed node, the relation (3) states that

$$f[x, x_0] = \frac{f(x) - f(x_0)}{x - x_0},$$

$$f[x, x_0, x_1] = \frac{f[x, x_0] - f[x_0, x_1]}{x - x_1},$$

etc. If we multiply these out and rearrange, we find that

$$f(x) = f(x_0) + (x - x_0)f[x, x_0],$$
$$f[x, x_0] = f[x_0, x_1] + (x - x_1)f[x, x_0, x_1],$$
$$f[x, x_0, x_1] = f[x_0, x_1, x_2] + (x - x_2)f[x, x_0, x_1, x_2],$$
$$\vdots$$
$$f[x, x_0, \ldots, x_{n-1}] = f[x_0, x_1, \ldots, x_n] + (x - x_n)f[x, x_0, x_1, \ldots, x_n].$$

Multiplying the second equation by $(x - x_0)$, the third by $(x - x_0)(x - x_1)$, and so forth, and adding the results yields

$$f(x) = P_n(x) + E_n(x) \tag{6a}$$

where

$$P_n(x) = f(x_0) + (x - x_0)f[x_0, x_1] + (x - x_0)(x - x_1)f[x_0, x_1, x_2] + \cdots + (x - x_0)(x - x_1)\cdots(x - x_{n-1})f[x_0, x_1, \ldots, x_n] \quad (6b)$$

and

$$E_n(x) = f[x, x_0, x_1, \ldots, x_n] \prod_{i=0}^{n} (x - x_i). \quad (6c)$$

$P_n(x)$ is the interpolating polynomial, and so we have established an error expression which we denote by $E_n(x)$:

$$\text{error at } x = f(x) - P_n(x) = E_n(x).$$

Let us look carefully at the error formula (6c). It is, of course, impossible to determine $f[x, x_0, \ldots, x_n]$ exactly, since the value $f(x)$ required for its evaluation is unknown. However, if an additional value of $f(x)$ is known, say $f(x_{n+1})$, and if we assume that the function $f[x, x_0, \ldots, x_n]$ is not changing rapidly on an interval containing x and $x_0, x_1, \ldots, x_{n+1}$, then

$$f[x, x_0, \ldots, x_n] \doteq f[x_{n+1}, x_0, x_1, \ldots, x_n].$$

This allows us to estimate the error by

$$E_n(x) \doteq f[x_0, \ldots, x_n, x_{n+1}] \prod_{i=0}^{n} (x - x_i). \quad (7)$$

To gain some theoretical insight into the meaning of the error expression (6c), we develop an alternative form. This second form supposes that $f(x)$ has $n + 1$ continuous derivatives; the differentiability is just needed in the derivation and is not actually necessary, as is clearly evidenced by the fact that (6c) is valid with f only continuous. We first write (6a) as

$$f(x) = P_n(x) + G(x) \prod_{i=0}^{n} (x - x_i) \quad (8)$$

where

$$G(x) = f[x, x_0, x_1, \ldots, x_n].$$

Fix x and define the function of t

$$S(t) = f(t) - P_n(t) - G(x) \prod_{i=0}^{n} (t - x_i). \quad (9)$$

If x is not equal to any of the x_i, we see that both

$$S(x) = 0$$

and

$$S(x_i) = 0, \qquad i = 0, 1, \ldots, n.$$

So the function $S(t)$ has $n + 2$ distinct zeros in the smallest interval I containing $x, x_0, x_1, \ldots, x_n$. Assuming suitable differentiability of $f(t)$, Rolle's theorem assures us that $S'(t)$ has at least $n + 1$ distinct zeros in I. Repeating the argument, we conclude that $S''(t)$ has at least n distinct zeros, $S'''(t)$ has at least $n - 1$ zeros, and finally $S^{(n+1)}(t)$ has at least one zero in I. Let z be one such zero. Differentiating equation (9) $n + 1$ times yields

$$S^{(n+1)}(t) = f^{(n+1)}(t) - P_n^{(n+1)}(t) - G(x)\frac{d^{n+1}}{dt^{n+1}}\left(\prod_{i=0}^{n}(t - x_i)\right).$$

$P_n(t)$ is a polynomial of degree n and so its $(n + 1)$st derivative vanishes. Since

$$\prod_{i=0}^{n}(t - x_i) = t^{n+1} + \text{lower degree terms},$$

its $(n + 1)$st derivative is simply $(n + 1)!$. So the expression for $S^{(n+1)}(t)$ simplifies to

$$S^{(n+1)}(t) = f^{(n+1)}(t) - G(x)(n + 1)!.$$

Evaluating this at $t = z$ and solving for $G(x)$, we have

$$G(x) = \frac{f^{(n+1)}(z)}{(n + 1)!},$$

where z is some (unknown, but dependent on x) point in I. The error expression (6c) may now be written as

$$E_n(x) = \frac{f^{(n+1)}(z)}{(n + 1)!}\prod_{i=0}^{n}(x - x_i). \tag{10}$$

This proof presumes that x is not a node, but the result is obviously true if x is a node for any choice of z in I and so is true for all x. If we define

$$M_{n+1} = \max_{t \in I} |f^{(n+1)}(t)|,$$

then an upper bound for $E_n(x)$ is given by

$$|E_n(x)| \leq \frac{M_{n+1}}{(n + 1)!}\prod_{i=0}^{n}|x - x_i|.$$

Example. Suppose we know $f(x) = \sin(x)$ at the nodes 0.0, 0.1, 0.2, 0.3, 0.4 and want to approximate $\sin(0.14)$ by $P_4(0.14)$. Since

$$M_5 = \max_{0.0 \leq t \leq 0.4} |\sin^{(5)}(t)| = \max_{0.0 \leq t \leq 0.4} |\cos(t)| = 1,$$

an upper bound for the error is given by

$$|E_4(0.14)| \leq \frac{1}{5!}|0.14 - 0.0| \times |0.14 - 0.1| \times |0.14 - 0.2| \times |0.14 - 0.3| \times |0.14 - 0.4| \doteq 1.2 \times 10^{-7}.$$

Example. If $f(x)$ is known at x_0 and x_1, linear interpolation gives the value

$$P_1(x) = a_0 + (x - x_0)a_1 = f(x_0) + (x - x_0)\frac{f(x_1) - f(x_0)}{x_1 - x_0}$$

and the error satisfies

$$f(x) - P_1(x) = \frac{f''(z)}{2}(x - x_0)(x - x_1).$$

If $x_0 < x < x_1$ and

$$\max_{x_0 \leq t \leq x_1} |f''(t)| = M_2,$$

then

$$|f(x) - P_1(x)| \leq M_2 \frac{|x - x_0|\,|x - x_1|}{2} \leq M_2 \frac{(x_1 - x_0)^2}{8}.$$

If we use linear interpolation on $f(x) = \sin x$ given the values at the nodes in the example above and $0.0 \leq x \leq 0.4$, then by using the nodes closest to x we can guarantee that the error is less than

$$1 \cdot \frac{(0.1)^2}{8} = 0.00125.$$

As a practical matter, since $f^{(n+1)}(x)$ is not known, the best that can be done to reduce the error (10) is to try to make the factor $\prod_{k=0}^{n}(x - x_k)$ small. As a rule of thumb, nodes $x_0, x_1, \ldots, x_n$ should be used which have x near the center of the interval spanned by the nodes. Extrapolation (x outside the interval spanned by the x_k) is a risky process as the factor $\prod_{k=0}^{n}(x - x_k)$ is apt to be large. It is not necessarily true that higher degree interpolating polynomials yield more accurate approximations. Even if the nodes x_k are equally spaced, using more nodes may produce less accurate results. An interesting example discussed in a problem is the function $f(x) = 1/(1 + 25x^2)$ for $-1 \leq x \leq 1$.

An Algorithm for Interpolation

When using formula (2), we saw that to obtain the next higher degree polynomial, we need only add a term involving a new divided difference of one higher order. All other terms remain unchanged. This observation allows us to design an efficient scheme for evaluating $P_n(x)$. The relationship (3), when interpreted properly, can be used to generate the required differences. If we make the correspondences in notation

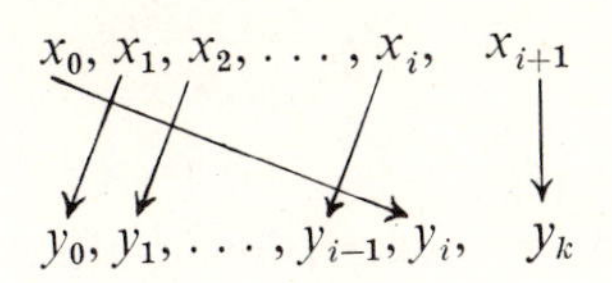

the relationship (3) for $f[x_0, x_1, \ldots, x_i, x_{i+1}]$ becomes

$$f[y_i, y_0, y_1, \ldots, y_{i-1}, y_k] = \frac{f[y_i, y_0, y_1, \ldots, y_{i-1}] - f[y_0, y_1, \ldots, y_{i-1}, y_k]}{y_i - y_k}$$

or, since the order of nodes is immaterial,

$$f[y_0, y_1, \ldots, y_i, y_k] = \frac{f[y_0, y\ , \ldots, y_i] - f[y_0, y_1, \ldots, y_{i-1}, y_k]}{y_i - y_k}. \tag{11}$$

Using (11) we can generate a table of divided differences as follows:

$$\begin{array}{llll}
x_0 \quad f_0 & & & \\
x_1 \quad f_1 & f_{01} = \dfrac{f_0 - f_1}{x_0 - x_1} & & \\
x_2 \quad f_2 & f_{02} = \dfrac{f_0 - f_2}{x_0 - x_2} & f_{012} = \dfrac{f_{01} - f_{02}}{x_1 - x_2} & \\
x_3 \quad f_3 & f_{03} = \dfrac{f_0 - f_3}{x_0 - x_3} & f_{013} = \dfrac{f_{01} - f_{03}}{x_1 - x_3} & f_{0123} = \dfrac{f_{012} - f_{013}}{x_2 - x_3},
\end{array}$$

etc. Note that each element in the table depends only on elements above and to the left of it. Because of this, the differences can be generated by rows from left to right, which allows us to compute only the differences that are actually required. The divided differences in the table above are used to evaluate $P_n(x)$ in the form (6b) as follows:

$$\pi_0 = 1, \qquad V_{0,0} = f_0, \qquad P_0(x) = V_{0,0}$$

$$\left.\begin{aligned}
V_{k,0} &= f_k \\
V_{k,i+1} &= \frac{V_{i,i} - V_{k,i}}{x_i - x_k}, \qquad i = 0, 1, \ldots, k-1 \\
\pi_k &= (x - x_{k-1})\pi_{k-1} \\
P_k(x) &= P_{k-1}(x) + \pi_k V_{k,k}
\end{aligned}\right\} \quad k = 1, 2, \ldots, n$$

where

$$V_{k,i} = f[x_0, x_1, \ldots, x_{i-1}, x_k], \qquad i = 1, 2, \ldots, k.$$

The expression

$$V_{k,i+1} = \frac{V_{i,i} - V_{k,i}}{x_i - x_k}$$

follows immediately from (11).

The important approximate expression for the error (7) shows that when interpolating with a polynomial of degree n using this scheme, the error is approximately the term to be added when going to degree $n + 1$. If one has a lot of data and seeks to interpolate to a specified accuracy, he can successively increase the number of interpolating points (and hence the degree of the polynomial) and obtain an estimate of the error in this way. If the term

$$\left| f[x_0, x_1, \ldots, x_{n+1}] \prod_{i=0}^{n} (x - x_i) \right|$$

becomes smaller than some specified error tolerance, one has estimated that the approximation using $P_n(x)$ is sufficiently accurate. Of course, one might as well use the better (ordinarily) result of $P_{n+1}(x)$ since the correction term to $P_n(x)$ is already available. We remind the reader of the practical observation that, when possible, using data roughly centered about x tends to give better results. Since the basic algorithm uses the data in the order $x_0, x_1, \ldots,$ the code supplied arranges the data so that x_0 is the node closest to x, and x_1 is the second closest, and so forth.

Differentiation

It is sometimes important to approximate the values of derivatives of $f(x)$. An obvious approach to this problem is to obtain the interpolating polynomial $P_n(x)$ and then approximate $f'(x)$ by $P_n'(x)$. One could use the same idea for the problem of higher derivatives; just approximate $f^{(k)}(x)$ by $P_n^{(k)}(x)$ if $k \leq n$. Unfortunately, considerable care is required since serious errors can arise in numerical differentiation. The inherent difficulty is that differentiation tends to magnify small discrepancies or errors in the approximating function $P_n(x)$; a typical situation is illustrated graphically in Figure II.1. Repeated differentiation tends to magnify these discrepancies still more. Because of this difficulty, numerical differentiation of the interpolating polynomial should be avoided whenever possible.

In the next section we discuss approximation by least squares polynomials. These polynomials tend to "smooth" the data and are often used to approximate derivatives. In the last section we introduce a different type of interpolating function, the cubic spline. Splines possess many interesting properties which make them useful for interpolation and differentiation.

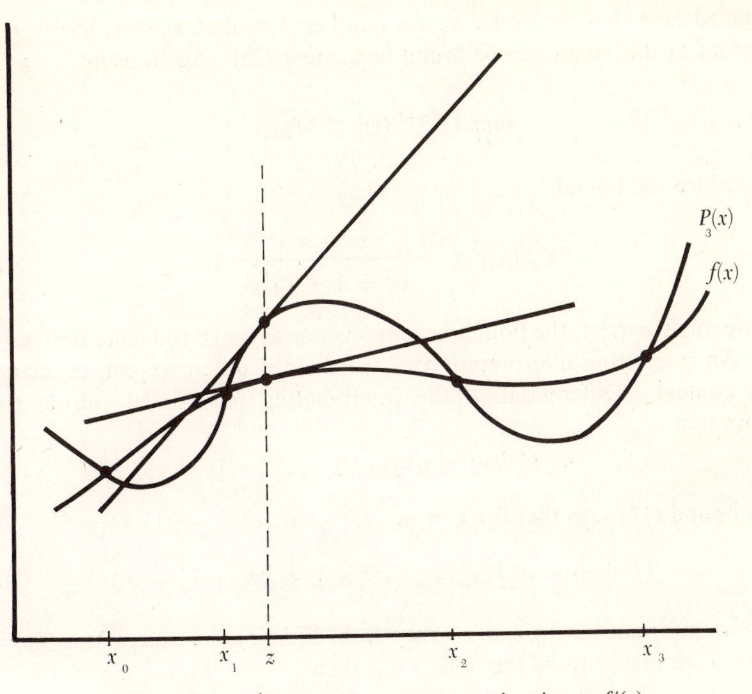

Figure II.1. $P_3'(z)$ may be a poor approximation to $f'(z)$

In spite of the problems associated with differentiation, the approximation of $f'(x)$ by $P_n'(x)$ is often used. Specific formulas will not be presented; all one has to do is to differentiate P_n as given by (6b). Instead we shall comment on the error. Recall that

$$f(x) = P_n(x) + E_n(x)$$

where P_n is given by (6b) and E_n by (6c) or (10). So,

$$f^{(k)}(x) = P_n^{(k)}(x) + E_n^{(k)}(x).$$

An argument very similar to that used in obtaining the expression (10) may be used to establish the result that for $k \leq n$

$$E_n^{(k)}(x) = \frac{f^{(n+1)}(z)}{(n+1-k)!} \prod_{j=0}^{n-k} (x - \zeta_j)$$

where the $n + 1 - k$ distinct points ζ_j are independent of x and are only known to lie in the intervals

$$x_j < \zeta_j < x_{j+k}, \qquad j = 0, 1, \ldots, n - k$$

(when $k = 0$, $\zeta_j = x_j$ for each j). The number z depends on x and is only known to be in the smallest interval I containing x and the ζ_j. We have assumed here that the nodes x_k are numbered so that $x_0 < x_1 < \cdots < x_n$. A proof of this result can be found in the text [26]. Again using

$$\max_{t \in I} |f^{(n+1)}(t)| \le M_{n+1},$$

we deduce the bound

$$|E_n^{(k)}(x)| \le \frac{M_{n+1}\,|x_0 - x_n|^{n+1-k}}{(n+1-k)!}. \tag{12}$$

As we might expect, the bound becomes worse as we go to higher derivatives.

An interesting interpretation of the divided difference can be made in this context. Differentiating the interpolating polynomial (6b) n times shows that

$$P_n^{(n)}(x) \equiv n!\,f[x_0, x_1, \ldots, x_n].$$

The bound (12) says that for $k = n$

$$|f^{(n)}(x) - n!\,f[x_0, x_1, \ldots, x_n]| \le M_{n+1}\,|x_0 - x_n|.$$

A more informative expression arises from equating the expressions (6c) and (10); if we take x to be the node x_{n+1}, then

$$f[x_0, x_1, \ldots, x_{n+1}] = \frac{f^{(n+1)}(z)}{(n+1)!}$$

and we see that the $(n+1)$st divided difference of $f(x)$ is related to the $(n+1)$st derivative of $f(x)$ evaluated at a suitable point.

It should be emphasized that up to now we have been assuming that errors in the function values do not play a dominant role. In the context of numerical differentiation of experimental data this is often invalid, and one must consider smoothing, as in the next section.

EXERCISES

1. Is the interpolating polynomial (6b) always of exact degree n? If not, illustrate by an example.

2. Suppose that $f(x)$ is a polynomial of degree n or less. Prove that if $P_n(x)$ interpolates $f(x)$ at $n+1$ distinct points, then $P_n(x) \equiv f(x)$. Make up an example and verify by direct calculation.

3. Consider the data:

x	$f(x)$
1	2
2	4

Construct $P_1(x)$ from (6b). Find polynomials $Q(x)$ and $S(x)$ of degree two and three respectively which interpolate the data. Does this contradict our theory?

4. Given the data

x	$f(x) = e^x - 50$
3.92	0.40044
3.94	1.41860
3.96	2.45733
3.98	3.51703
4.00	4.59815
4.02	5.70117

use the algorithm to approximate $f(3.947)$ using polynomials of degree $n = 1, 2, 3$. For each value of n use two different sets of nodes; for example, for $n = 2$ one might use the nodes 3.92, 3.94, 4.02 or 3.94, 3.96, 4.00. In each case estimate the error $E_n(3.947)$ using the approximation

$$\text{a)} \quad f[x_0, x_1, \ldots, x_n, x_{n+1}] \prod_{i=0}^{n} (x - x_i)$$

and the bound

$$\text{b)} \quad \frac{\max_{t \in I} |f^{(n+1)}(t)|}{(n+1)!} \prod_{i=0}^{n} |x - x_i|.$$

Compare with the actual errors.

5. The error function $\text{erf}(z)$ is defined by the integral

$$\text{erf}(z) = \frac{2}{\sqrt{\pi}} \int_0^z e^{-t^2}\, dt;$$

it is encountered frequently in the theories of probability, errors of observation, refraction, conduction of heat, etc. The function $\text{erf}(z)$ cannot be expressed in terms of the more familiar functions and hence must be evaluated numerically. Using the data

z	$\text{erf}(z)$
0.00	0.0000000
0.05	0.0563720
0.10	0.1124629
0.15	0.1679960
0.20	0.2227026

and the code INTPOL, approximate $\text{erf}(0.14)$ by polynomials of degree $n = 1, 2$, and 3. In each case print out the error approximation (7). Compare with the value $\text{erf}(0.14)$ which, correct to 7 decimal places, is 0.1569470. Suppose one were to do linear interpolation in the above table. What is the worst error that could be made?

6. Given $n + 1$ distinct points $x_0, x_1, \ldots, x_n$, recall that there is a unique polynomial $P_n(x)$ interpolating f at these points. There are many forms of

$P_n(x)$ which are different from the one that we presented in the text. For example, the Lagrangian form is

$$P_n(x) = \sum_{k=0}^{n} f(x_k)L_k(x)$$

where

$$L_k(x) = \prod_{\substack{m=0 \\ m \neq k}}^{n} \frac{x - x_m}{x_k - x_m}.$$

For this form verify that
a) Each $L_k(x)$ is a polynomial of degree $\leq n$; hence $P_n(x)$ is of degree $\leq n$.
b) For $x = x_k$, L_k has the special value $L_k(x_k) = 1$; if $k \neq j$, $L_k(x_j) = 0$.
c) $P_n(x)$ interpolates $f(x)$ on the nodes $x_0, x_1, \ldots, x_n$.
What are some advantages (or disadvantages) of this form with respect to the form (6b)? For some interesting comparisons, see [28].

7. In performing potentiometric titrations one obtains a potential difference curve plotted against volume of titrant added. The *equivalence* point is defined to be the inflection point (point at which the second derivative vanishes) of the curve. Some instruments electronically differentiate this curve twice and obtain the equivalence point as the zero of the second derivative. Without such a device it is necessary to do this numerically. The following table is taken from A. I. Vogel, *A Textbook of Quantitative Inorganic Analysis*, 3rd ed., John Wiley & Sons, New York, 1961, p. 930. It gives the measurements for the potentiometric titration of Fe^{2+} solution with 0.1095 N Ce^{4+} solution using platinum and calomel electrodes.

Solution added (ml)	E (mV)
1.00	373.00
5.00	415.00
10.00	438.00
15.00	459.00
20.00	491.00
21.00	503.00
22.00	523.00
22.50	543.00
22.60	550.00
22.70	557.00
22.80	565.00
22.90	575.00
23.00	590.00
23.10	620.00
23.20	860.00
23.30	915.00
23.40	944.00
23.50	958.00
24.00	986.00
26.00	1067.00
30.00	1125.00

Sketch the graph of E as a function of solution added. Examination of the curve shows it to be relatively flat at both ends of the data with a steep rise near the equivalence point. Such a curve is difficult to approximate by polynomials—the more so the steeper the rise—which leads to poor approximation of the derivatives. This is compensated, however, by the fact that a steep rise means that the equivalence point is located in a small region.

Write out an explicit expression for the second derivative of the quadratic polynomial interpolating on three successive points. Apply it to the above problem of finding an approximation for the equivalence point. This is actually the way chemists do it.

8. Crank and Park, Trans. Faraday Soc. *45*(1949), 240–249, have devised a method of deducing the diffusion coefficient D for chloroform in polystyrene from uptake measurements. Using several assumptions, they arrive at the quantity

$$\bar{D}(C_0) = \frac{1}{C_0}\int_0^{C_0} D(C)\,dC$$

which can be measured for a number of C_0 values. A differentiation with respect to C_0 gives an expression for D in terms of the quantity $\frac{d}{dC_0}[C_0\bar{D}(C_0)]$. As only a crude approximation is reasonable at this stage of the process, use linear or quadratic interpolation and the data

C_0	$\bar{D}(C_0)$
5.0	0.0240
7.5	0.0437
9.9	0.0797
12.9	0.1710
13.2	0.1990
15.1	0.3260
16.3	0.8460
16.8	0.9720

to approximate D for each C_0 value.

9. In the text it was shown that for $n + 1$ distinct nodes $x_0, x_1, \ldots, x_n$ there is one and only one polynomial of degree n, $P_n(x)$, which has given values f_k at each node x_k. In general this is not true if the degree is $m \neq n$. If $m < n$, argue that a subset of the data uniquely defines $P_m(x)$ and that in general not all the interpolation conditions can be satisfied. If the degree is $m > n$, there are many polynomials that interpolate. To see this we observe that there is always a polynomial $Q(x)$ of degree n which interpolates. If $R(x)$ is *any* polynomial of degree $m - n - 1$, then

$$P(x) = Q(x) + R(x)\prod_{k=0}^{n}(x - x_k)$$

is an interpolating polynomial of degree m. Verify.

10. Interpolation frequently becomes awkward in the neighborhood of a singularity of a function or of one of its derivatives. A commonly used technique is to introduce an auxiliary function either additively or multiplicatively. Thus, we might choose functions $s(x)$ or $p(x)$ which are simpler in the sense of permitting

"easier" interpolation in such a way that the functions $S(x) = s(x) + f(x)$ or $P(x) = f(x)p(x)$ are smoother than $f(x)$ itself.

As an example, consider the problem of interpolating $\csc(x)$ near $x = 0.0$ using the following data:

x	$\csc(x)$
0.000	∞
0.005	200.0010
0.010	100.0020
0.015	66.6692
0.020	50.0033
0.030	33.3383

The Laurent series for $\csc(x)$,

$$\csc(x) = \frac{1}{x} + \frac{1}{6}x + \frac{7}{360}x^3 + \cdots$$

suggests that the functions

$$x\csc(x) \quad \text{and} \quad \csc(x) - \frac{1}{x}$$

are well behaved near $x = 0.0$. Approximate $\csc(0.001)$ by using the above data. For the same nodes construct tables of $\csc(x) - \frac{1}{x}$ and $x\csc(x)$. Interpolate in these tables at $x = 0.001$ and hence evaluate $\csc(0.001)$. Compare results.

11. As indicated in the text, the function $f(x) = 1/(1 + 25x^2)$ on $-1 \leq x \leq 1$ is an example for which, if one uses equally spaced nodes, higher degree interpolating polynomials do not necessarily yield more accurate approximations. Defining the nodes

$$x_i = -1 + ih, \qquad i = 0, 1, \ldots, n, \qquad h = \frac{2}{n},$$

use the code INTPOL to generate data so that you may sketch the graphs of $P_5(x)$ and $P_{20}(x)$ for $-1 \leq x \leq 1$. Compare with $f(x)$. You will need to alter the dimensions in INTPOL to compute $P_{20}(x)$.

12. The potential energy of two or more interacting molecules is called Van der Waal's interaction energy. A theoretical calculation for two interacting helium atoms has produced the set of energies $V(r)$ for various values of the internuclear distance r given on p. 43. The energy exhibits repulsion ($V > 0$) for small r and attraction ($V < 0$) for larger values of r. In addition, there is a minimum corresponding to stable equilibrium ($V' = 0$) at $r = r_e$ and an inflection point ($V'' = 0$) at $r = r_i$.

    Approximate r_e and r_i by differentiating a quadratic interpolating polynomial.

r (bohrs)	$V(r)$ (°K)
4.6	32.11
4.8	9.00
5.0	−3.52
5.1	−7.11
5.2	−9.22
5.3	−10.74
5.4	−11.57
5.5	−11.95
5.6	−12.00
5.7	−11.73
5.8	−11.21
5.9	−10.71
6.0	−10.13
6.5	−7.15
7.0	−4.77
7.5	−3.17
8.0	−2.14
9.0	−1.03
10.0	−0.54

13. Problem 11 shows that high order interpolation does not necessarily give good approximations even when the function is smooth and the nodes are spread evenly. Piecewise polynomial interpolation will, however. For example, suppose we use the nodes

$$a = x_0 < x_1 < \cdots < x_n = b$$

and in each subinterval $[x_i, x_{i+1}]$ the additional nodes $x_{i,0}$, $x_{i,1}$ with $x_i < x_{i,0} < x_{i,1} < x_{i+1}$. The piecewise cubic polynomial function $S(x)$ interpolating to $f(x)$ on these nodes is defined by: for $i = 0, 1, \ldots, n-1$, $S(x)$ is defined on $[x_i, x_{i+1}]$ as the cubic polynomial interpolating to $f(x)$ at $x_i, x_{i,0}, x_{i,1}, x_{i+1}$.
    Prove that $S(x) \in C[a, b]$. Suppose that $f(x) \in C^4[a, b]$ and let $\max |f^{(4)}(x)| = M_4$, $\max(x_{i+1} - x_i) = h$. Using inequality (12), prove that for $x \in [a, b]$,

$$|f^{(k)}(x) - S^{(k)}(x)| \leq \frac{M_4}{(4-k)!} h^{4-k} \qquad k = 0, 1, 2.$$

This result is intimately related to the cubic splines studied in II.1.3.

1.2 *LEAST SQUARES APPROXIMATION

In the preceding section we sought the polynomial of degree n which agreed with given function values $f(x_j)$ at $n + 1$ distinct nodes $x_1, x_2, \ldots, x_n, x_{n+1}$. There are circumstances in which it is appropriate to generalize this task. One such is data compression. Suppose that we have data $f(x_1), f(x_2), \ldots, f(x_m)$ which we seek to approximate by a polynomial for analytical purposes, or because it is more convenient than a table. If m

is large, the degree of the interpolating polynomial is also large and the polynomial is itself inconvenient to deal with; we need a lower degree polynomial approximation. The second circumstance is closely related. Suppose the data $f(x_j)$ are subject to error, for example, errors of measurement. We might be quite certain that the underlying function $f(x)$ is essentially a polynomial of low degree n, but the errors in the data cause the various polynomials of degree n interpolating on different subsets of the data to differ enormously. It seems reasonable that if one uses a "lot" of measurements, then the effect of the errors in the data could be averaged out. This feeling can be put on a sound statistical foundation, but for our purposes it will suffice to suggest that it is reasonable to seek a (the?) polynomial of degree n, $P_n(x)$, which minimizes the root mean square (RMS) error of the approximation:

$$\left(\frac{1}{m}\sum_{j=1}^{m}(P_n(x_j) - f(x_j))^2\right)^{1/2}.$$

If there are $m \leq n + 1$ pieces of data, we can interpolate to get a polynomial which gives the minimum value of zero; the interesting case is when $m > n + 1$. It may happen that some measurements $f(x_j)$ are more reliable than others and we want $P_n(x)$ to agree with them more closely. This case is included by introducing weights $w_j > 0$ which measure the relative "goodness" of the $f(x_j)$ and by seeking a polynomial $P_n(x)$ which minimizes

$$\sum_{j=1}^{m} w_j(P_n(x_j) - f(x_j))^2. \tag{1}$$

Let us denote this minimum value by $m\epsilon_n^2$. Another reason one might want weights is that without them we are minimizing an average absolute error. To make this a weighted relative error for $f(x_j) \neq 0$ with weights $\mu_j > 0$, simply take

$$w_j = \frac{\mu_j}{f(x_j)^2};$$

for then we are minimizing

$$\sum_{j=1}^{m} \mu_j\left(\frac{P_n(x_j) - f(x_j)}{f(x_j)}\right)^2.$$

The approach we use to solve this problem is related mathematically to Fourier analysis. The underlying mathematical structure is both elegant and illuminating, but we forego its development to keep the mathematical prerequisites to a minimum. This unfortunately makes the development quite formal, though it is not difficult. The reader familiar with Fourier series will see many analogies.

We use the notation (u, v), as applied to two functions $u(x)$, $v(x)$ defined on the set $\{x_1, x_2, \ldots, x_m\}$, to mean

$$(u, v) = \sum_{j=1}^{m} w_j u(x_j) v(x_j). \tag{2}$$

There are several properties the reader can now prove very easily. For any functions $u(x)$ and $v(x)$,

$$(u, u) \geq 0,$$

$$(u, u) = 0 \quad \text{means} \quad u(x_j) = 0 \quad \text{for all } j,$$

and

$$(u, v) = (v, u).$$

For any (real) constants $\alpha_1, \ldots, \alpha_p$ and functions $v_1(x), v_2(x), \ldots, v_p(x)$ defined on $\{x_1, \ldots, x_m\}$,

$$\left(u, \sum_{i=1}^{p} \alpha_i v_i\right) = \sum_{i=1}^{p} \alpha_i (u, v_i).$$

We are going to construct a set of polynomials $Q_0(x), \ldots, Q_k(x), \ldots, Q_n(x)$ such that for each k, $Q_k(x)$ is of exact degree k and furthermore

$$(Q_k, Q_j) = 0 \quad \text{if} \quad k \neq j. \tag{3}$$

(By exact degree k, we mean $Q_k(x) = \gamma x^k +$ lower powers of x and $\gamma \neq 0$.) These properties will make it easy to solve the minimum problem. Indeed, assuming that we can construct such polynomials, we shall now prove that the unique polynomial minimizing (1) is

$$R_n(x) = \sum_{k=0}^{n} c_k Q_k(x), \tag{4}$$

where

$$c_k = (Q_k, f)/(Q_k, Q_k). \tag{5}$$

First we show this definition makes sense. The denominator of each c_k satisfies

$$(Q_k, Q_k) \geq 0;$$

if it vanishes, then $Q_k(x_j) = 0$ for each $j = 1, \ldots, m$. Hence $Q_k(x)$, a polynomial of degree $k < m$, has m distinct real roots; as we argued in II.1.1 this is a contradiction which implies $(Q_k, Q_k) > 0$.

In our notation (2) the sum (1) is just $(P_n - f, P_n - f)$. Using the rules above, it can be written as

$$(P_n - f, P_n - f) = (P_n - R_n + R_n - f, P_n - R_n + R_n - f)$$
$$= (P_n - R_n, P_n - R_n) + 2(P_n - R_n, R_n - f) + (R_n - f, R_n - f). \quad (6)$$

We leave as an exercise the proof of the fact that any polynomial of degree n can be written as a linear combination of $Q_0(x), \ldots, Q_n(x)$; so let

$$P_n(x) - R_n(x) = \sum_{k=0}^{n} \alpha_k Q_k(x).$$

Then

$$(P_n - R_n, R_n - f) = \sum_{k=0}^{n} \alpha_k (Q_k, R_n - f).$$

Note that each term satisfies

$$(Q_k, R_n - f) = \sum_{j=0}^{n} c_j (Q_k, Q_j) - (Q_k, f)$$
$$= c_k (Q_k, Q_k) - (Q_k, f) = 0$$

because of (3), (4), and (5). Thus, from (6) we see that

$$(P_n - f, P_n - f) = (P_n - R_n, P_n - R_n)$$
$$+ (R_n - f, R_n - f) \geq (R_n - f, R_n - f).$$

This says that the polynomial $R_n(x)$ gives the smallest possible value of $(P_n - f, P_n - f)$. If P_n gives the same value, then

$$(P_n - R_n, P_n - R_n) = 0$$

and so

$$P_n(x_j) - R_n(x_j) = 0 \quad \text{for} \quad j = 1, \ldots, m.$$

Considering the degrees of the polynomials involved, this implies that $P_n(x) - R_n(x) \equiv 0$; hence $R_n(x)$ is the unique minimizing polynomial.

If we can generate the $Q_k(x)$ fairly easily, we have a simple solution to the least squares approximation problem. This turns out to be the case, and there are still other nice properties. The minimum error is

$$(R_n - f, R_n - f) = (R_n, R_n) - 2(R_n, f) + (f, f).$$

Expanding the terms shows

$$(R_n, R_n) = \sum_{i=0}^{n} \sum_{j=0}^{n} c_i c_j (Q_i, Q_j) = \sum_{k=0}^{n} c_k^2 (Q_k, Q_k)$$
$$= \sum_{k=0}^{n} c_k (Q_k, f).$$

Since

$$(R_n, f) = \sum_{k=0}^{n} c_k(Q_k, f),$$

we infer from (1) that

$$m\epsilon_n^2 = \sum_{j=1}^{m} w_j(R(x_j) - f(x_j))^2 = \sum_{j=1}^{m} w_j f^2(x_j) - \sum_{k=0}^{n} c_k^2(Q_k, Q_k).$$

Examining what we have done, it is clear that the best polynomial of any degree $s < m$ is

$$R_s(x) = \sum_{k=0}^{s} c_k Q_k(x) \tag{7}$$

and its error is

$$m\epsilon_s^2 = \sum_{j=1}^{m} w_j f^2(x_j) - \sum_{k=0}^{s} c_k^2(Q_k, Q_k). \tag{8}$$

Thus, by computing the best polynomial of degree N in this way we also find the best polynomials of degrees $0, 1, \ldots, N - 1$ at the same time. This is extremely valuable when one seeks to balance the inconvenience of a high degree polynomial against the fact that it gives a smaller error. Indeed, since the $Q_k(x)$ are generated in the order $k = 0, 1, \ldots, N$, it is not necessary to decide in advance what degree to use. The code supplied tries to meet a specified error tolerance by using successively higher degree polynomials subject to a maximum degree specified by the user.

The polynomials $Q_k(x)$ can be defined recursively as follows:

$$Q_0(x) \equiv 1,$$

$$Q_1(x) = x - a_1,$$

$$Q_k(x) = (x - a_k)Q_{k-1}(x) - b_k Q_{k-2}(x) \qquad k = 2, 3, \ldots,$$

where

$$a_k = (xQ_{k-1}, Q_{k-1})/(Q_{k-1}, Q_{k-1}) \qquad k = 1, 2, \ldots, \tag{9a}$$

$$b_k = (xQ_{k-1}, Q_{k-2})/(Q_{k-2}, Q_{k-2}) \qquad k = 2, 3, \ldots. \tag{9b}$$

We shall first prove that the $Q_k(x)$ are well defined and of exact degree k. This is evident for $k = 0$. For $k = 1$,

$$a_1 = (x, 1)/(1, 1) = \sum_{j=1}^{m} w_j x_j \Big/ \sum_{j=1}^{m} w_j$$

is well defined and the degree of $Q_1(x)$ is obvious. Now we argue by induction. If we assume that $Q_{k-2}(x)$ and $Q_{k-1}(x)$ are of exact degree $k - 2$ and $k - 1$

respectively, the denominators in the definitions of a_k and b_k are positive, so these quantities are well defined. The definition

$$Q_k(x) = xQ_{k-1}(x) - a_kQ_{k-1}(x) - b_kQ_{k-2}(x)$$

clearly makes $Q_k(x)$ of exact degree k.

The proof of the property

$$(Q_i, Q_j) = 0 \quad \text{if} \quad i \neq j$$

is equivalent to showing that

$$(Q_k, Q_i) = 0 \quad \text{if} \quad i < k$$

for each k. We prove this by induction on k. If $k = 0$, there is nothing to prove. If $k = 1$,

$$\begin{aligned}(Q_1, Q_0) &= (x - a_1, 1) = (x, 1) - a_1(1, 1) \\ &= 0\end{aligned}$$

by the definition of a_1. Suppose the result is true for $0, 1, \ldots, k - 1$. Then

$$\begin{aligned}(Q_k, Q_{k-1}) &= (xQ_{k-1}, Q_{k-1}) - a_k(Q_{k-1}, Q_{k-1}) - b_k(Q_{k-1}, Q_{k-2}) \\ &= (xQ_{k-1}, Q_{k-1}) - a_k(Q_{k-1}, Q_{k-1}) \\ &= 0,\end{aligned}$$

using the induction assumption and the definition of a_k. In the same way

$$(Q_k, Q_{k-2}) = 0$$

from the definition of b_k. Lastly, if $i < k - 2$,

$$\begin{aligned}(Q_k, Q_i) &= (xQ_{k-1}, Q_i) - a_k(Q_{k-1}, Q_i) - b_k(Q_{k-1}, Q_i) \\ &= (xQ_{k-1}, Q_i) = (Q_{k-1}, xQ_i) \\ &= (Q_{k-1}, Q_{i+1}) + a_{i+1}(Q_{k-1}, Q_i) + b_{i+1}(Q_{k-1}, Q_{i-1}) \\ &= 0.\end{aligned}$$

The reader is invited to verify this last identity in detail.

The recursion furnishes a simple computational algorithm for the $Q_k(x)$. It can be made still more simple by using the identity

$$b_k = (Q_{k-1}, Q_{k-1})/(Q_{k-2}, Q_{k-2}). \tag{9b'}$$

To establish this, comparison with (9b) shows that we must prove

$$(xQ_{k-1}, Q_{k-2}) = (Q_{k-1}, Q_{k-1}).$$

The left side is

$$(xQ_{k-1}, Q_{k-2}) = (Q_{k-1}, xQ_{k-2}),$$

and the right side is equal to this same quantity, since

$$\begin{aligned}(Q_{k-1}, Q_{k-1}) &= (Q_{k-1}, xQ_{k-2} - a_{k-1}Q_{k-2} - b_{k-1}Q_{k-3}) \\ &= (Q_{k-1}, xQ_{k-2}) - a_{k-1}(Q_{k-1}, Q_{k-2}) - b_{k-1}(Q_{k-1}, Q_{k-3}) \\ &= (Q_{k-1}, xQ_{k-2}).\end{aligned}$$

We have found the best least squares approximation in the form

$$R_n(x) = \sum_{k=0}^{n} c_k Q_k(x).$$

For some purposes one might want to express this in the power form

$$R_n(x) = \sum_{k=0}^{n} e_k x^k.$$

This can be done in a straightforward manner using the recurrence relation for the $Q_k(x)$, but should be done in multiple precision since the transformation can be numerically delicate. The use of nested multiplication to evaluate the power form can be written as the recurrence

$$\begin{cases} d_{n+1} = 0 \\ \quad d_k = xd_{k+1} + e_k \qquad k = n, n-1, \ldots, 0 \\ R_n(x) = d_0. \end{cases}$$

This is a cheap way to evaluate the power form, but this whole process can be moderately inaccurate. It is frequently more accurate to use the form involving the $Q_k(x)$ directly. It can be evaluated by the recurrence

$$\begin{cases} d_{n+2} = d_{n+1} = 0 \\ \quad d_k = c_k + (x - a_{k+1})d_{k+1} - b_{k+2}d_{k+2} \qquad k = n, n-1, \ldots, 0 \\ R_n(x) = d_0. \end{cases} \tag{10}$$

It is easy to prove that this works. First we solve for c_k in this relation and substitute it in the expression for $R_n(x)$:

$$\begin{aligned}R_n(x) &= \sum_{k=0}^{n} c_k Q_k(x) = \sum_{k=0}^{n} [d_k - (x - a_{k+1})d_{k+1} + b_{k+2}d_{k+2}]Q_k(x) \\ &= \sum_{k=0}^{n} d_k Q_k(x) - \sum_{k=0}^{n} d_{k+1}(x - a_{k+1})Q_k(x) + \sum_{k=0}^{n} d_{k+2}b_{k+2}Q_k(x).\end{aligned}$$

Now change the indices of summation and recombine to obtain

$$R_n(x) = d_0Q_0(x) + d_1Q_1(x) + \sum_{k=2}^{n} d_kQ_k(x)$$

$$- d_1(x - a_1)Q_0(x) - \sum_{k=2}^{n} d_k(x - a_k)Q_{k-1}(x) + \sum_{k=2}^{n} d_kb_kQ_{k-2}(x)$$

$$= d_0Q_0(x) + d_1[Q_1(x) - (x - a_1)Q_0(x)]$$

$$+ \sum_{k=2}^{n} d_k[Q_k(x) - (x - a_k)Q_{k-1}(x) + b_kQ_{k-2}(x)],$$

where we use the fact that $d_{n+1} = d_{n+2} = 0$. Now, using the recurrence relation for the $Q_k(x)$ and the fact that $Q_0(x) \equiv 1$, we see that

$$R_n(x) = d_0.$$

This way of evaluating $R_n(x)$ is more expensive than nested multiplication for the power form but it is often more accurate. We suggest that it be used unless one is genuinely interested in the power form. In the latter case the transformation to the power form should definitely be done using multiple precision.

The recurrence (10) was written as though the d_k were constants because we had in mind evaluating it for a specific number x. But it holds for each x so that one can properly think of $d_k(x)$ as a polynomial defined by

$$\begin{cases} d_{n+2}(x) = d_{n+1}(x) \equiv 0 \\ d_k(x) = c_k + (x - a_{k+1})d_{k+1}(x) - b_{k+2}d_{k+2}(x) \qquad k = n, n-1, \ldots, 0 \\ R_n(x) = d_0(x). \end{cases} \tag{10'}$$

This way of regarding the recurrence is the key to developing recurrences for the derivatives of $R_n(x)$ and to converting $R_n(x)$ to power form.

The implementation of the computational algorithm in the code LEAST is not entirely straightforward. The basic computational step is the formation of inner products such as

$$(u, v) = \sum_{j=1}^{m} w_ju(x_j)v(x_j).$$

As we discussed in part I, Floating Point Arithmetic, the judicious use of double precision can greatly enhance the accuracy of inner products, especially since in practical fitting the number of terms may be rather large. Some inner products, e.g.,

$$(u, u) = \sum_{j=1}^{m} w_ju^2(x_j),$$

do not require this double precision, so the code LEAST accumulates inner products only when it is advantageous.

The code monitors the RMS error of the various fits as it raises the degree, but it does not use the convenient expression (8). Errors in forming (8), even in double precision, can lead to *negative* values of $m\epsilon_s^2$. So we work directly with

$$m\epsilon_s^2 = \sum_{j=1}^{m} w_j(R_s(x_j) - f(x_j))^2,$$

which does not have this flaw. The values $R_s(x_j)$ are formed as in (7), but accumulating the inner product. Frequently these values of the fit are the most interesting to the user, so we are partially compensated for the expense of obtaining an accurate measure of the error when it is small.

A point which is not obvious is that we do not use the expression (5) to compute c_k. Rather, we use the equivalent form

$$c_k = (Q_k, f - R_{k-1})/(Q_k, Q_k). \tag{11}$$

This is equivalent because

$$(Q_k, f - R_{k-1}) = (Q_k, f) - \sum_{j=0}^{k-1} c_j(Q_k, Q_j) = (Q_k, f).$$

This form is better behaved numerically because we are calculating c_k so that the expression

$$R_{k-1}(x) + c_k Q_k(x)$$

approximates $f(x)$ as well as it can. The form (11) chooses c_k so that the term added, $c_k Q_k(x)$, reduces the error $f(x) - R_{k-1}(x)$ as much as is possible. The point is that it is reducing the error of the *computed* approximation $R_{k-1}(x)$ rather than the theoretical one which leads to the first form. Since we carry double precision values for $R_{k-1}(x_j)$ along, it costs practically nothing for us to use the better form. In an exercise we ask the reader to verify the difference in performance of the two forms.

EXERCISES

1. In the text we used the fact that if $Q_0(x), \ldots, Q_n(x)$ are polynomials such that $Q_k(x)$ is of exact degree k for each k, then any polynomial $P_n(x)$ of degree n can be written in the form

$$P_n(x) = \sum_{k=0}^{n} \alpha_k Q_k(x).$$

Prove this by induction on the degree of the polynomial to be represented. What must α_n be so that $P_n(x) - \alpha_n Q_n(x)$ is a polynomial of degree $n - 1$? The induction assumption is that any polynomial of degree $n - 1$ can be represented as a linear combination of $Q_0(x), \ldots, Q_{n-1}(x)$.

2. Least squares fits are often used for numerical differentiation; devise algorithms for evaluating

$$R_n^{(p)}(x) = \sum_{k=0}^{n} c_k Q_k^{(p)}(x)$$

directly for $p = 1, 2, \ldots$.

3. T. W. Murphy in "Modeling of Lung Gas Exchange—Mathematical Models of the Lung," Mathematical Biosciences *5* (1969), 427–447, develops a computer simulation of lung gas exchange. He requires a relationship between carbon dioxide (CO_2) content in the liquid phase and the CO_2 partial pressure in the gas phase. Develop an empirical equation by fitting a cubic polynomial to the following data:

Partial Pressure (mm Hg)	Concentration (cc CO_2/cc blood)
20	0.368
30	0.432
40	0.485
50	0.528
60	0.567

Considering the errors of the fit at the data points, does a cubic seem adequate?

4. Write out explicit formulas for a least squares fit of degree 1 to $f(x_j)$ at $\{x_1, \ldots, x_m\}$ in terms of the quantities $(1, 1)$, $(1, x)$, $(1, f)$, (x, x), (x, f). This is an extremely useful special case of our analysis.

5. The first quantitative study of the velocity or rate of a chemical reaction was performed by Wilhelmy in 1850 (Pogg. Ann. *81*, 413) on the "inversion" of cane sugar (sucrose) in acid solution. The experimental quantity observed was the optical rotation, α, of the sucrose as a function of time, t. Since optical rotation depends linearly on the concentration of sucrose in water, C, he was able to establish that the rate of reaction, defined to be $-\frac{dC}{dt}$, was "first order in sucrose." This means that $-\frac{dC}{dt} = kC$, where k is a proportionality constant. A student has obtained the following data in an attempt to reproduce Wilhelmy's results:

t (minutes)	rotation $\alpha(t)$ (degrees)
0	21.65
5	21.40
10	21.20
15	21.00
20	20.65
25	20.50
30	20.20
40	19.80
50	19.25
60	18.90
75	18.15
90	17.75
105	16.95
120	16.55
135	16.10
150	15.65
∞	4.25

Estimate the proportionality constant k.

A standard method of solution is as follows: Since C depends linearly on α, we may write

$$\frac{C(t) - C(0)}{C(0) - C(\infty)} = \frac{\alpha(t) - \alpha(0)}{\alpha(0) - \alpha(\infty)}.$$

From the differential equation satisfied by C, $C(t) = C(0)e^{-kt}$, the above equation becomes

$$e^{-kt} = \frac{\alpha(t) - \alpha(\infty)}{\alpha(0) - \alpha(\infty)}$$

Taking logarithms of both sides gives

$$\ln(\alpha(t) - \alpha(\infty)) = -kt + \ln(\alpha(0) - \alpha(\infty)).$$

This is automatically true for $t = 0$; for $t \neq 0$ we have

$$k = -\frac{1}{t}\ln\left(\frac{\alpha(t) - \alpha(\infty)}{\alpha(0) - \alpha(\infty)}\right).$$

For the data gathered by the student, form the function of t on the right hand side of the last equation and determine k by fitting a constant to this function. If your constant value is a good fit, it should change little as the degree of the fit is raised. Using LEAST and EVAL, fit the function with polynomials of degrees 0 through 5 and determine their constant terms. You can do this with one call to LEAST and the fact that the constant term of $R_n(t)$ is just $R_n(0)$.

6. Prove that

a) $(u, u) \geq 0$,
b) $(u, u) = 0$ means $u(x_j) = 0$ for all j,
c) $(u, v) = (v, u)$
and for constants $\alpha_1, \alpha_2, \ldots, \alpha_p$
d) $\left(u, \sum_{i=1}^{p} \alpha_i v_i\right) = \sum_{i=1}^{p} \alpha_i(u, v_i)$.

7. It is very easy to alter the code LEAST to use the expression (5) to compute the coefficient c_k. Just change the statement

```
TEMP = DBLE(F(I)) - R(I)
```

to

```
TEMP = DBLE(F(I))
```

in the two places it appears.

Compare the two procedures by fitting the data $x_i = i$, $f(x_i) = x_i^7$ for $i = 1, 2, \ldots, 20$. Since the fit is theoretically exact for degree 7, the RMS error should be zero for degrees 7, 8, ... Using weights $w_i = 1/f(x_i)^2$ (that is, pure relative error) and the two codes, calculate the RMS errors for degrees 7, 8, 9, and 10. Which form is better computationally?

1.3 *CUBIC INTERPOLATORY SPLINES

Often a person with some data to interpolate has some feeling on physical grounds as to the smoothness of the underlying function. Interpolating polynomials may, when plotted, show oscillations which are physically unreasonable. This often happens when there are sudden changes in an otherwise slowly varying function. Draftsmen draw smooth curves through such data points by using splines. These are thin flexible strips of plastic or other material which are laid on the graph paper and held with weights so that the spline must go over data points. The weights are so constructed that the spline is free to slip, and as a result the flexible spline straightens out as much as it can subject to passing over the data points. The draftsman then traces along the spline to get his interpolating curve. The cubic interpolatory spline we discuss here is a mathematical model of the physical spline which allows us to mechanize the process. Although we shall be interested in the mathematical spline because of its own good properties, it is true that it closely approximates the physical spline when curvatures are small.

Suppose the data are $f(x_0), f(x_1), \ldots, f(x_n)$ with the nodes ordered $a = x_0 < x_1 < \cdots < x_n = b$. Nodes are often called "joints" in this context. The natural cubic interpolatory spline $S(x)$, or as we shall say for brevity, the cubic spline, is defined by the following properties:

(i) $S(x)$ is continuous along with its first and second derivatives on $[a, b]$.
(ii) $S(x_i) = f_i, \qquad i = 0, 1, \ldots, n.$
(iii) $S(x)$ is a cubic polynomial on each interval $[x_i, x_{i+1}], \qquad i = 0, 1, \ldots, n-1.$
(iv) $S''(x_0) = 0, \qquad S''(x_n) = 0.$

Notice that in contrast to polynomial interpolation, which increases the degree to interpolate at more points, here the degree is fixed and one uses more polynomials instead. It is more work to calculate a spline and more trouble to use because we have a collection of polynomials, but splines are a very satisfactory device for smooth interpolation and numerical differentiation.

The first order of business is to show that the properties listed actually define a function $S(x)$ uniquely. We shall do this by showing how to construct it in a way which makes a very good algorithm. For notational convenience, let

$$h_i = x_{i+1} - x_i,$$

$$S_i(x) \equiv S(x), \qquad x \in [x_i, x_{i+1}],$$

$$s_i = S''(x_i).$$

Of course, we do not yet know what the s_i are. As we successively require $S(x)$ to satisfy the four properties (i–iv), we find equations for the s_i. Later we shall show how to solve these equations, which not only proves that $S(x)$ exists, but also shows how to compute it. Because $S_i(x)$ is a cubic polynomial, $S_i''(x)$ is a linear polynomial and can be expressed in the form

$$S_i''(x) = s_i \frac{x_{i+1} - x}{h_i} + s_{i+1} \frac{x - x_i}{h_i}, \qquad i = 0, 1, \ldots, n-1. \tag{1}$$

Clearly an $S''(x)$ so defined on the subintervals is a continuous function of x on $[a, b]$. To obtain expressions for the $S_i(x)$, integrate (1) twice to obtain

$$S_i(x) = \frac{s_i}{6h_i}(x_{i+1} - x)^3 + \frac{s_{i+1}}{6h_i}(x - x_i)^3 + c_1(x - x_i) + c_2(x_{i+1} - x). \tag{2}$$

The constants of integration c_1, c_2 may be determined from the interpolation condition $S_i(x_i) = f_i$, $S_i(x_{i+1}) = f_{i+1}$. So,

$$S_i(x_i) = f_i = \frac{s_i}{6} h_i^2 + c_2 h_i,$$

$$S_i(x_{i+1}) = f_{i+1} = \frac{s_{i+1}}{6} h_i^2 + c_1 h_i$$

lead to the expressions

$$c_1 = \frac{f_{i+1}}{h_i} - \frac{s_{i+1}h_i}{6}, \qquad c_2 = \frac{f_i}{h_i} - \frac{s_i h_i}{6}.$$

Equations (2) then become

$$S_i(x) = \frac{s_i}{6h_i}(x_{i+1} - x)^3 + \frac{s_{i+1}}{6h_i}(x - x_i)^3 + \left(\frac{f_{i+1}}{h_i} - \frac{s_{i+1}h_i}{6}\right)(x - x_i)$$
$$+ \left(\frac{f_i}{h_i} - \frac{s_i h_i}{6}\right)(x_{i+1} - x), \qquad i = 0, 1, \ldots, n-1. \tag{3}$$

The interpolation condition will make the function $S(x)$ represented by (3) continuous on $[a, b]$. Differentiating (3) yields

$$S_i'(x) = -\frac{s_i}{2h_i}(x_{i+1} - x)^2 + \frac{s_{i+1}}{2h_i}(x - x_i)^2 + \frac{f_{i+1} - f_i}{h_i} - \frac{h_i}{6}(s_{i+1} - s_i). \tag{4}$$

To get a function with $S'(x)$ continuous requires that

$$S_{i-1}'(x_i) = S_i'(x_i).$$

Using (4), we must then have

$$s_{i+1} + 2s_i\left(\frac{h_i + h_{i-1}}{h_i}\right) + \frac{h_{i-1}}{h_i}s_{i-1}$$
$$= \frac{6}{h_i}\left(\frac{f_{i+1} - f_i}{h_i} - \frac{f_i - f_{i-1}}{h_{i-1}}\right), \qquad i = 1, \ldots, n-1. \quad (5)$$

The set of equations (5) is a system of $n - 1$ linear equations in the $n + 1$ unknowns $s_0, s_1, \ldots, s_n$. If we can find a solution, then we shall have produced an $S(x)$ with the properties (i), (ii), and (iii). Two additional conditions are needed to specify a unique solution. The requirement (iv), which is $s_0 = 0$, $s_n = 0$, leads to what is called the natural spline. This corresponds to letting the physical spline stick out past the end weights. Usually we shall not be interested in $x < a$ or $x > b$ but if they arise, the natural spline is extended by straight lines agreeing in value and slope at the ends a and b. (Note that the second derivative is then also continuous at the ends.) There are other possibilities that we shall not discuss, but we should mention that if one has the data $f'(x_0), f'(x_n)$ available, another good choice is to use

(iv′) $$S'(x_0) = f'(x_0), \qquad S'(x_n) = f'(x_n).$$

Here we just work with

$$s_0 = 0, \qquad s_n = 0. \quad (6)$$

The method we use to solve the system (5), (6) is a special case of the general method of part III, although it looks rather different. Once we solve this system of equations, the spline is specified by the expressions (3). Consider (5) written in the form

$$\frac{h_{i-1}}{h_i}s_{i-1} + 2\left(1 + \frac{h_{i-1}}{h_i}\right)s_i + s_{i+1} = d_i \qquad i = 1, 2, \ldots, n-1 \quad (7)$$

along with (6). Here

$$d_i = \frac{6}{h_i}\left(\frac{f_{i+1} - f_i}{h_i} - \frac{f_i - f_{i-1}}{h_{i-1}}\right).$$

Let us define

$$s_{i-1} = \rho_i s_i + \tau_i, \qquad i = 1, 2, \ldots, n, \quad (8)$$

where we seek suitable quantities ρ_i and τ_i. Since $s_0 = 0$ we can take $\rho_1 = 0$ and $\tau_1 = 0$. Substitution of (8) into (7) and a little manipulation shows

$$s_i = \frac{-1}{\dfrac{h_{i-1}}{h_i}\rho_i + 2\left(1 + \dfrac{h_{i-1}}{h_i}\right)}s_{i+1} + \frac{d_i - \dfrac{h_{i-1}}{h_i}\tau_i}{\dfrac{h_{i-1}}{h_i}\rho_i + 2\left(1 + \dfrac{h_{i-1}}{h_i}\right)}.$$

This has the same form as (8) and suggests defining

$$\rho_{i+1} = \frac{-1}{\dfrac{h_{i-1}}{h_i}\rho_i + 2\left(1 + \dfrac{h_{i-1}}{h_i}\right)}, \qquad \tau_{i+1} = \frac{d_i - \dfrac{h_{i-1}}{h_i}\tau_i}{\dfrac{h_{i-1}}{h_i}\rho_i + 2\left(1 + \dfrac{h_{i-1}}{h_i}\right)}. \tag{9}$$

If we can prove that no denominator vanishes, this gives a simple recursion for computing ρ_i and τ_i for $i = 1, 2, \ldots, n$. Then, starting with $s_n = 0$, we can use (8) to calculate the s_i in the order $n - 1, n - 2, \ldots, 1$.

On proving that

$$\frac{h_{i-1}}{h_i}\rho_i + 2\left(1 + \frac{h_{i-1}}{h_i}\right) \neq 0 \quad \text{for all} \quad i, \tag{10}$$

we will have completed the proof that the cubic spline exists and is unique, and will have developed a simple algorithm for computing it. Clearly, $|\rho_1| < 1$. If $|\rho_i| < 1$, we have

$$\begin{aligned}\left|\frac{h_{i-1}}{h_i}\rho_i + 2\left(1 + \frac{h_{i-1}}{h_i}\right)\right| &= \left|\frac{h_{i-1}}{h_i}(\rho_i + 2) + 2\right| \\ &\geq \left|\frac{h_{i-1}}{h_i} + 2\right| > 1\end{aligned}$$

so that (10) holds. Moreover, this inequality and the definition (9) show that $|\rho_{i+1}| < 1$. But then (10) holds for the case $i + 1$, etc.

The spline has some important properties which make it useful for interpolation and differentiation. First we prove the minimum "curvature" property and demonstrate a reason for the choice $s_0 = s_n = 0$. Suppose $f''(x)$ is continuous on $[a, b]$ and $S(x)$ is the spline interpolating to $f(x)$ on $\{x_0, x_1, \ldots, x_n\}$. Let $g(x)$ be any other function with two continuous derivatives interpolating $f(x)$ on $\{x_0, x_1, \ldots, x_n\}$ (for example, $P_n(x)$, the interpolating polynomial). Then a bit of manipulation shows that

$$\begin{aligned}\int_a^b [g''(t) - S''(t)]^2\,dt = \int_a^b [g''(t)]^2\,dt& \\ - 2\int_a^b [g''(t) - S''(t)]S''(t)\,dt& \\ - \int_a^b [S''(t)]^2\,dt.&\end{aligned} \tag{11}$$

We shall prove that the second integral on the right hand side of (11) is zero. Recalling that $a = x_0 < x_1 < \cdots < x_n = b$, this integral can be written as

$$\sum_{i=0}^{n-1} \int_{x_i}^{x_{i+1}} S''(t)[g''(t) - S''(t)]\, dt.$$

An integration by parts gives

$$\sum_{i=0}^{n-1} \left\{ [(g'(t) - S'(t))S''(t)] \Big|_{x_i}^{x_{i+1}} - \int_{x_i}^{x_{i+1}} [g'(t) - S'(t)]S'''(t)\, dt \right\}.$$

Note that on $[x_i, x_{i+1}]$, $S'''(t)$ is constant; call this constant α_i. Then we can integrate

$$\int_{x_i}^{x_{i+1}} [g'(t) - S'(t)]S'''(t)\, dt = \alpha_i [g(t) - S(t)] \Big|_{x_i}^{x_{i+1}}$$

which vanishes because both g and S agree with f at each joint, i.e.,

$$g(x_i) - S(x_i) = f(x_i) - f(x_i) = 0.$$

We are then left with

$$\sum_{i=0}^{n-1} \{S''(x_{i+1})[g'(x_{i+1}) - S'(x_{i+1})] - S''(x_i)[g'(x_i) - S'(x_i)]\}.$$

This sum telescopes to

$$S''(x_n)[g'(x_n) - S'(x_n)] - S''(x_0)[g'(x_0) - S'(x_0)] \tag{12}$$

which is also zero since we have assumed that

$$S''(x_n) = s_n = 0 = s_0 = S''(x_0).$$

Equation (11) may now be written in the form

$$\int_a^b [g''(t)]^2\, dt = \int_a^b [S''(t)]^2\, dt + \int_a^b [g''(t) - S''(t)]^2\, dt.$$

We remark that if $f(x)$ has a continuous second derivative, it can be taken as $g(x)$ here to arrive at

$$\int_a^b [f''(t)]^2\, dt = \int_a^b [S''(t)]^2\, dt + \int_a^b [f''(t) - S''(t)]^2\, dt.$$

This will be useful in our study of the error of interpolation. Returning now to the general case, we see that

$$\int_a^b [g''(t)]^2\, dt \geq \int_a^b [S''(t)]^2\, dt. \tag{13}$$

Thus, among all twice-continuously-differentiable interpolating functions, the spline with $s_0 = s_n = 0$ has the least value of this integral. In this sense the spline fit provides the smoothest interpolating function to f. The integrand is approximately the curvature when the curvature is small. Another way to view this result is that a straight line would give the minimum value of zero to the integral and in effect $S(x)$ is as "close" to a straight line as we can get subject to the requirement that we interpolate.

It should be pointed out that if equality holds in (13), we must have

$$g(x) \equiv S(x).$$

This follows by noting that $(g'' - S'')^2$ is continuous and so

$$\int_a^b (g'' - S'')^2\, dx = 0$$

implies

$$S''(x) = g''(x).$$

Integrating twice, we obtain

$$S(x) = g(x) + c_1 x + c_2,$$

where c_1 and c_2 are constants. Since S and g both interpolate $f(x)$ at x_0 and x_1, we have

$$f_0 = f_0 + c_1 x_0 + c_2$$

$$f_1 = f_1 + c_1 x_1 + c_2$$

which implies, since $x_0 \neq x_1$,

$$c_1 = c_2 = 0.$$

So $S(x)$ gives the unique minimum of

$$\int_a^b [g''(x)]^2\, dx$$

subject to the conditions stated above.

Another point worth noting is that the minimum "curvature" property is also valid if, instead of assuming $s_0 = 0 = s_n$, we ask that

$$S'(x_0) = f'(x_0), \qquad S'(x_n) = f'(x_n)$$

and, of course, that $g(x)$ satisfy the same conditions. This result follows immediately since these conditions also make the expression (12) vanish. The remainder of the argument is unchanged.

Next we shall obtain bounds on the quantities

$$|f(x) - S(x)|, \qquad |f'(x) - S'(x)|, \qquad x \in [a, b]$$

and establish the following theorem.

Theorem 1. Let f be twice continuously differentiable on $[a, b]$ and $S(x)$ be the natural cubic spline interpolating $f(x)$ at $a = x_0 < x_1 < \cdots < x_n = b$. If we define

$$h = \max_{0 \le i \le n-1} (x_{i+1} - x_i),$$

then

$$\max_{a \le x \le b} |f(x) - S(x)| \le h^{3/2} \left\{ \int_a^b [f''(t)]^2 \, dt \right\}^{1/2}, \tag{14}$$

$$\max_{a \le x \le b} |f'(x) - S'(x)| \le h^{1/2} \left\{ \int_a^b [f''(t)]^2 \, dt \right\}^{1/2}. \tag{15}$$

PROOF. Let x be any point in the interval $[a, b]$. For some i, $x \in [x_i, x_{i+1}]$. Since $f(t) - S(t)$ vanishes at x_i and x_{i+1}, Rolle's theorem states that there is a point z in $[x_i, x_{i+1}]$ where

$$f'(z) - S'(z) = 0.$$

Then

$$\int_z^x [f''(t) - S''(t)] \, dt = [f'(t) - S'(t)] \Big|_z^x = f'(x) - S'(x).$$

The Cauchy-Schwarz inequality [38] implies that

$$\begin{aligned} |f'(x) - S'(x)| &= \left| \int_z^x [f''(t) - S''(t)] \cdot 1 \, dt \right| \\ &\le \left| \int_z^x [f''(t) - S''(t)]^2 \, dt \right|^{1/2} \left| \int_z^x 1^2 \, dt \right|^{1/2} \\ &= \left| \int_z^x [f''(t) - S''(t)]^2 \, dt \right|^{1/2} |x - z|^{1/2} \\ &\le \left| \int_z^x [f''(t) - S''(t)]^2 \, dt \right|^{1/2} h^{1/2}. \end{aligned}$$

As we have already remarked,

$$\int_a^b [f''(t) - S''(t)]^2\,dt = \int_a^b [f''(t)]^2\,dt - \int_a^b [S''(t)]^2\,dt$$
$$\leq \int_a^b [f''(t)]^2\,dt.$$

Then since the interval $[z, x]$ is contained in $[a, b]$,

$$|f'(x) - S'(x)| \leq h^{1/2}\left\{\int_a^b [f''(t)]^2\,dt\right\}^{1/2}$$

for all $x \in [a, b]$. This establishes (15).

Again let $x \in [a, b]$; there is an i such that $x \in [x_i, x_{i+1}]$ and

$$f(x) - S(x) = \int_{x_i}^x [f'(t) - S'(t)]\,dt.$$

So

$$|f(x) - S(x)| = \left|\int_{x_i}^x [f'(t) - S'(t)]\,dt\right|$$
$$\leq \int_{x_i}^x |f'(t) - S'(t)|\,dt$$
$$\leq \int_{x_i}^x \max_{a \leq z \leq b} |f'(z) - S'(z)|\,dt$$
$$\leq h \max_{a \leq z \leq b} |f'(z) - S'(z)|.$$

Using (15), we see that

$$|f(x) - S(x)| \leq h^{3/2}\left\{\int_a^b [f''(t)]^2\,dt\right\}^{1/2}.$$

Theorem 1 tells us that as the number of nodes is increased (assuming of course that this decreases h), S' and S converge uniformly to f' and f respectively for all $x \in [a, b]$. With the "smoothness" property of S, this makes the spline a useful tool for interpolation and a particularly useful tool for numerical differentiation.

Often one can assert that convergence is faster than this theorem states because we employ a simple, elementary proof in this theorem. A useful example of a stronger result presumes that $f(x) \in C^4[a, b]$ and that as $h \to 0$ the mesh satisfies

$$\max_i (h/h_i) \leq \beta < \infty$$

for a fixed β. Then

$$\max_{a \leq x \leq b} |f^{(k)}(x) - S^{(k)}(x)| = 0(h^{4-k}), \qquad k = 0, 1, 2.$$

EXERCISES

1. Consider the function $f(x) = 1/(1 + 25x^2)$ on $-1 \leq x \leq 1$. For the equally spaced nodes

$$x_i = -1 + ih, \qquad i = 0, 1, \ldots, n, \qquad h = \frac{2}{n},$$

use the codes provided and generate data to plot the cubic spline fit to $f(x)$ for $n = 5$ and 20. Compare with the results obtained in problem 11 of II.1.1.

2. A common problem is to approximate

$$\int_a^b f(x)\,dx$$

when $f(x)$ is known only at the points $x_0 < x_1 < \cdots < x_n$. One approach is to compute the cubic spline $S(x)$ and to use

$$\int_a^b S(x)\,dx$$

as an approximation. Suppose $x_0 = a$ and $x_n = b$; derive a specific formula for

$$\int_{x_0}^{x_n} S(x)\,dx.$$

Write a FORTRAN subroutine to be used with SPCOEF to evaluate the integral of the spline.

3. Even when accurate values for a function $f(x)$ are known at the points $x_0 < x_1 < \cdots < x_n$, one often finds differentiating an interpolating polynomial a highly unsatisfactory way to approximate $f'(x)$. Roughly speaking, interpolating polynomials tend to have "ripples." Theorem 1 and the physical view of spline interpolation suggest that a good way to approximate $f'(x)$ is to compute the cubic interpolatory spline $S(x)$ and to use $S'(x)$. Using the formulas developed in the text, write a FORTRAN subroutine to be used with SPCOEF to evaluate $S'(x)$.

4. Develop a quadratic interpolating spline $S(x)$. With nodes $x_0 < x_1 < \cdots < x_n$ and $f_i = f(x_i)$ given at each x_i, make $S(x)$ satisfy:

   (i) $S(x)$ is continuous along with its first derivative on $[x_0, x_n]$;
   (ii) $S(x_i) = f_i$, $i = 0, 1, \ldots, n$;
   (iii) $S(x)$ is a quadratic polynomial on each interval $[x_i, x_{i+1}]$, $i = 0, 1, \ldots, n - 1$.

To make your work simpler, write

$$S(x) = a_i + b_i(x - x_i) + c_i(x - x_i)(x - x_{i+1}) \quad \text{for} \quad x_i \leq x \leq x_{i+1}.$$

Derive explicit expressions for the a_i and b_i from property (ii). Using property (i), prove that

$$c_i = \frac{b_i - b_{i-1}}{x_{i+1} - x_i} - c_{i-1}\frac{x_i - x_{i-1}}{x_{i+1} - x_i}, \qquad i = 1, 2, \ldots, n - 1.$$

Just as we saw with the cubic splines, the properties (i, ii, iii) do not completely determine the spline. Argue that if one chooses a value for any one of the c_i, then all the rest are easily calculated and the spline determined. Noting that $S''(x) = 2c_i$ for $x_i < x < x_{i+1}$, how could you use divided differences to choose a suitable value for some c_i?

5. Chemicals appear colored as a result of absorbing certain wavelengths of light from incident white light (which contains all wavelengths, or colors), the remaining colors being transmitted. The amount of light absorbed (absorbance) as a function of wavelength is called the spectrum of the chemical. In a research experiment the spectrum of a solution of vanadyl D-tartrate dimer was obtained. Part of the results are given below. By differentiating the spline fit, find the wavelength which yields the maximum absorbance.

Wavelength (Angstroms)	Absorbance
5000.	0.350
4875.	0.252
4750.	0.183
4625.	0.144
4500.	0.170
4375.	0.315
4250.	0.530
4125.	0.750
4000.	0.836
3875.	0.769
3750.	0.560
3625.	0.449
3500.	0.382
3375.	0.400
3250.	0.572
3125.	0.700

2. INTEGRATION

Interpolatory Methods

In this chapter we shall be concerned with the problem of approximating the integral

$$\int_a^b f(x)\,dx \tag{1}$$

when a and b are finite. The basic idea is to approximate $f(x)$ by an interpolating polynomial and then integrate the polynomial in place of $f(x)$. Methods of this type are called interpolatory. A simple example is based on the straight line interpolating $f(x)$ at the end points of the interval. The polynomial is

$$P(x) = f(a) + (x - a)\,\frac{f(b) - f(a)}{b - a},$$

and performing the integration results in

$$\int_a^b f(x)\,dx \doteq \frac{b - a}{2}\,[f(a) + f(b)], \tag{2}$$

the trapezoidal rule. Notice that the interpolating polynomial disappears on integration and we are just left with a linear combination of function values to approximate the integral.

A natural way to get a better approximation than the trapezoidal rule is to use an interpolating polynomial of higher degree. Using a quadratic polynomial interpolating at a, $\frac{a+b}{2}$, and b gives a rule called Simpson's rule. The polynomial is

$$P(x) = f(a) + (x - a)f[a, \tfrac{a+b}{2}] + \left(x - \frac{a+b}{2}\right)(x - a)f[a, \tfrac{a+b}{2}, b]$$

and the resulting rule is

$$\int_a^b f(x)\,dx \doteq S(f) = \frac{b-a}{6}\left[f(a) + 4f\left(\frac{a+b}{2}\right) + f(b)\right]. \tag{3}$$

This rule has some nice features and will be the basis of all our subsequent analysis. Of course, there are many other possibilities for a basic formula arising from the various choices of degree and nodes, but it appears that efficient integration depends more on how one uses an effective basic formula than on the formula itself

Error of Simpson's Rule

Let us now study the accuracy of the basic Simpson's rule (3). The error expressions, or rather their derivations, are unmotivated but quite simple. They just involve repeated integration by parts. First define $c = \dfrac{a+b}{2}$ and split the integral into two pieces:

$$\int_a^b f(x)\,dx = \int_a^c f(x)\,dx + \int_c^b f(x)\,dx.$$

In general, integration by parts gives

$$\int_a^c f(x)\,dx = f(x)(x-\alpha_1)\Big|_a^c - \int_a^c f'(x)(x-\alpha_1)\,dx,$$
$$\int_c^b f(x)\,dx = f(x)(x-\alpha_2)\Big|_c^b - \int_c^b f'(x)(x-\alpha_2)\,dx$$

where α_1, α_2 are constants which we are free to choose. Adding these expressions gives the identity

$$\int_a^b f(x)\,dx = f(b)(b-\alpha_2) + f(c)(\alpha_2-\alpha_1) + f(a)(\alpha_1-a)$$
$$- \int_a^c f'(x)(x-\alpha_1)\,dx - \int_c^b f'(x)(x-\alpha_2)\,dx.$$

Now we choose α_1 and α_2 so as to get Simpson's rule:

$$b-\alpha_2 = \frac{b-a}{6},$$

$$\alpha_2-\alpha_1 = \frac{4(b-a)}{6},$$

$$\alpha_1-a = \frac{b-a}{6}.$$

There are three equations and only two unknowns, but there is the unique solution

$$\alpha_1 = \frac{b+5a}{6}, \qquad \alpha_2 = \frac{5b+a}{6}.$$

Defining the function $u(x)$ by

$$u(x) = \begin{cases} x - \alpha_1 & \text{if} \quad a \le x \le c \\ x - \alpha_2 & \text{if} \quad c < x \le b, \end{cases}$$

we have

$$\int_a^b f(x)\,dx = S(f) - \int_a^b f'(x)u(x)\,dx. \tag{4}$$

We formulate this as a theorem.

Theorem 1. If $f(x)$ has a continuous derivative on $[a, b]$, then (4) holds. Furthermore if

$$M_1 \ge |f'(x)| \quad \text{for} \quad a \le x \le b,$$

then

$$\left| \int_a^b f(x)\,dx - S(f) \right| \le M_1 \int_a^b |u(x)|\,dx.$$

If $f(x)$ has more continuous derivatives, the same device of appropriately choosing the constants in the integration by parts can be used several more times. It is straightforward to prove the following result.

Theorem 2. If $f(x)$ has four continuous derivatives, then

$$\int_a^b f(x)\,dx = S(f) + \int_a^b f^{(4)}(x)v(x)\,dx \tag{5}$$

where

$$v(x) = \begin{cases} \dfrac{(x-a)^3}{24}\left(x - \dfrac{a+2b}{3}\right) & \text{if} \quad a \le x \le c \\ \dfrac{(x-b)^3}{24}\left(x - \dfrac{2a+b}{3}\right) & \text{if} \quad c < x \le b. \end{cases}$$

Furthermore, there is a point ξ with $a < \xi < b$ such that

$$\int_a^b f(x)\,dx = S(f) - \frac{(b-a)^5}{2880} f^{(4)}(\xi). \tag{6}$$

Only (6) requires comment. Because $v(x)$ is negative on (a, b), we can use an integral mean value theorem to write

$$\int_a^b f^{(4)}(x)v(x)\,dx = f^{(4)}(\xi) \int_a^b v(x)\,dx$$

and we simply calculate

$$\int_a^b v(x)\,dx = -\frac{(b-a)^5}{2880}.$$

These results suggest that one would rarely apply Simpson's rule directly to approximate (1) since it is not very accurate. The way we get a good approximation is to observe that the length of the interval $|b - a|$ appears in the error expressions raised to a power. If we can reduce this length a little, we shall reduce the error a lot. This paradoxical task is easily accomplished by splitting up the original problem. For example, we might choose an integer N and split

$$\int_a^b f(x)\,dx = \sum_{i=0}^{N-1} \int_{a+i(b-a)/N}^{a+(i+1)(b-a)/N} f(x)\,dx.$$

Applying Simpson's rule to each integral leads to

$$\int_a^b f(x)\,dx \doteq \sum_{i=0}^{N-1} \frac{b-a}{6N}\left[f\left(a + \frac{i(b-a)}{N}\right) + 4f\left(a + \frac{(i+\frac{1}{2})(b-a)}{N}\right) + f\left(a + \frac{(i+1)(b-a)}{N}\right)\right].$$

This formula, which we denote by $S_N(f)$, is called the compound, composite, or repeated Simpson's rule. If we let $h = \dfrac{b-a}{N}$ and collect terms, we have

$$S_N(f) = \frac{h}{6}\left[f(a) + 4\sum_{j=0}^{N-1} f(a + (j + \tfrac{1}{2})h) + 2\sum_{j=1}^{N-1} f(a + jh) + f(b)\right].$$

Iterative Procedures

We ask rather more of an integration code than simply returning a value alleged to be close to the desired integral. The code is expected to assess its error and attempt to produce a value satisfying some accuracy criterion specified by the user. Naturally, we expect very accurate results to cost more than moderately accurate results, and the user must balance his desire for accuracy against the cost of attaining it. An iterative procedure computes a sequence of approximate integrals of increasing accuracy, assesses their accuracy, and terminates with the best value when it is adjudged to have met the accuracy requirement.

One way to construct an iterative code is based on the compound Simpson's rule. To assess the error, suppose we have the simple approximation to the basic integral,

$$\int_a^b f(x)\,dx \doteq S_1(f).$$

Split the interval $[a, b]$ into several pieces; splitting in half is referred to as bisection, in three pieces as trisection, and the like. We shall always use bisection. Thus, we consider

$$\int_a^b f(x)\,dx \doteq S_2(f).$$

On simple grounds of consistency we might hope that the error in $S_2(f)$ will be less than the difference $|S_1(f) - S_2(f)|$; when $f(x)$ is smooth, we can be more precise. From Theorem 2,

$$\int_a^b f(x)\,dx = S_1(f) - \frac{(b-a)^5}{2880} f^{(4)}(\xi_1)$$

for $a < \xi_1 < b$. Applying this also to each half interval, we have

$$\int_a^b f(x)\,dx = \int_a^{(a+b)/2} f(x)\,dx + \int_{(a+b)/2}^b f(x)\,dx$$

$$= S_2(f) - \frac{(b-a)^5}{2880} \cdot \frac{1}{2^5} \cdot [f^{(4)}(\xi_{21}) + f^{(4)}(\xi_{22})]$$

where $a < \xi_{21} < \dfrac{a+b}{2} < \xi_{22} < b$. We now make the crucial assumption that the interval $[a, b]$ is small enough so that $f^{(4)}(x)$ is essentially constant; we denote this constant by $f^{(4)}$. Then

$$\int_a^b f(x)\,dx \doteq S_1(f) - \frac{(b-a)^5}{2880} f^{(4)},$$

$$\int_a^b f(x)\,dx \doteq S_2(f) - \frac{(b-a)^5}{2880} f^{(4)} \cdot \frac{1}{2^4}.$$

Subtraction yields

$$S_2(f) - S_1(f) \doteq -\frac{2^4 - 1}{2^4} \frac{(b-a)^5}{2880} f^{(4)}$$

from which it easily follows that

$$\int_a^b f(x)\,dx - S_2(f) \doteq \frac{1}{2^4 - 1} [S_2(f) - S_1(f)]. \tag{7}$$

The relationship (7) furnishes us with a method for assessing the error of an approximate value of the integral. The compound Simpson's rule $S_N(f)$ splits the interval $[a, b]$ into the N equal subintervals $[a + (i(b-a)/N), a + ((i+1)(b-a)/N)]$, $i = 0, 1, \ldots, N-1$ and applies Simpson's rule to each subinterval. The rule $S_{2N}(f)$ just amounts to splitting each of these N subintervals into two parts and applying Simpson's rule to each part. The approximate error expression (7) then allows us to estimate the error over each of the N subintervals. If we add up all of the approximate integrals as well as the error terms, we arrive at

$$\int_a^b f(x)\,dx - S_{2N}(f) \doteq \tfrac{1}{15}[S_{2N}(f) - S_N(f)].$$

Suppose we seek an approximate integral $Q(f)$ such that

$$\left| \int_a^b f(x)\,dx - Q(f) \right| \leq \epsilon.$$

We simply compute the sequence of approximations $S_{2^k}(f)$ for $k = 0, 1, \ldots$ and monitor the difference

$$|S_{2^{k+1}}(f) - S_{2^k}(f)|.$$

If this difference becomes less than ϵ, then we exit with $S_{2^{k+1}}(f)$ as $Q(f)$. Notice that we actually estimate the error to be less than $\epsilon/15$; the fraction helps compensate for the crudeness of the error estimate, especially when k is small. One might well prefer to use the value

$$Q(f) = S_{2^{k+1}}(f) + \tfrac{1}{15}[S_{2^{k+1}}(f) - S_{2^k}(f)]$$

which subtracts out the estimated error.

Recall that our crucial assumption in deriving the error expression was that $f^{(4)}(x)$ be essentially constant over the intervals $[a + (i(b - a)/N), a + ((i + 1)(b - a)/N)]$. If $f^{(4)}(x)$ is continuous and if we take $N = 2^k$ sufficiently large, this is certainly valid. With any iterative procedure of the kind derived here, one is implicitly making an assumption of this sort. He must always consider the possibility that a code terminates because the assumption is not valid rather than because the result is accurate. There is a natural tendency to be suspicious when N is quite small at termination, although of course it is possible that the problem is just "easy." Adding in the estimated error also depends on this basic assumption, so it is most likely to be valid when ϵ is "small" or k is "large."

An Adaptive Procedure

The procedures we have been discussing can be quite inefficient in one respect. We split $[a, b]$ into many pieces and compute two approximations to the integral over each piece. If the agreement is not satisfactory, we split again. The inefficiency arises because, although the agreement may be satisfactory on many (perhaps nearly all) subintervals, a large disagreement in just one interval forces refinement in all. The idea of an adaptive code is to refine only where necessary.

Our adaptive procedure is defined as follows. Compute

$$Q_1(f) = (b-a)\cdot\tfrac{1}{6}\cdot\left[f(a) + 4f\left(\frac{a+b}{2}\right) + f(b)\right]$$

and

$$Q_{11}(f) = \left(\frac{b-a}{2}\right)\cdot\tfrac{1}{6}\cdot\left[f(a) + 4f\left(a+\frac{b-a}{4}\right) + f\left(\frac{a+b}{2}\right)\right],$$

$$Q_{12}(f) = \left(\frac{b-a}{2}\right)\cdot\tfrac{1}{6}\cdot\left[f\left(\frac{a+b}{2}\right) + 4f\left(a+\frac{3(b-a)}{4}\right) + f(b)\right]$$

and compare Q_1 to $Q_{11} + Q_{12}$. Is

$$|Q_1 - (Q_{11} + Q_{12})| \leq \epsilon?$$

If so, report $Q_{11} + Q_{12}$ as the answer. If not, set the right half interval $\left[\frac{a+b}{2}, b\right]$ aside for the moment and proceed in the same way for the left half interval as for the original one. We already have

$$Q_{11} \doteq \int_a^{(a+b)/2} f(x)\,dx$$

and now we compute

$$Q_{111} = \left(\frac{b-a}{2^2}\right)\cdot\tfrac{1}{6}\cdot\left[f(a) + 4f\left(a+\frac{b-a}{8}\right) + f\left(a+\frac{b-a}{4}\right)\right],$$

$$Q_{112} = \left(\frac{b-a}{2^2}\right)\cdot\tfrac{1}{6}\cdot\left[f\left(a+\frac{b-a}{4}\right) + 4f\left(a+\frac{3(b-a)}{8}\right) + f\left(\frac{a+b}{2}\right)\right].$$

Notice that no function evaluations from the previous step are wasted. We now ask whether

$$|Q_{11} - (Q_{111} + Q_{112})| \leq \epsilon/2. \tag{8}$$

If not, we set the right half interval $\left[a + \frac{b-a}{4}, \frac{a+b}{2}\right]$ aside for the moment and repeat the process for the left half interval. We shall speak of levels of refinement, with Q_1 being from level one, Q_{11} and Q_{12} from level two, Q_{111} and Q_{112} from level 3, and so forth. If the test (8) is passed, we accept

$$Q_{111} + Q_{112} \doteq \int_a^{(a+b)/2} f(x)\,dx$$

and proceed up one level and to the right. So, we now improve Q_{12} by computing Q_{121} and Q_{122}. If

$$|Q_{12} - (Q_{121} + Q_{122})| \leq \epsilon/2,$$

we accept $Q_{121} + Q_{122}$, add it to the approximation previously obtained and move to the right. Normally one proceeds to the right until an approximation has been found for the whole interval $[a, b]$.

It is a practical necessity to limit the number of levels used because the number of function evaluations increases rapidly as we proceed to higher levels. On reaching the maximum level permitted, one could accept the last approximation computed and move to the right or terminate in a suitable way. The accuracy of the function values $f(x)$ clearly affects how accurately integrals of $f(x)$ can be computed. We naturally presume that the original problem has a value of the integral which is large compared to the errors in the function values. As we proceed to higher levels, the lengths of the intervals, and consequently the magnitudes of the corresponding integrals over these intervals, decrease so that errors in computing $f(x)$ become more and more important in the sense of relative error. Ultimately the errors in $f(x)$ become limiting and further refinement can be detrimental.

This adaptive process refines often where the integral of $f(x)$ is difficult to approximate and uses few levels where it is easy. The device can greatly increase the efficiency of a quadrature code. Though the basic idea is simple, it can be a little tricky to code the process so as not to evaluate the integrand more than once at any x. The code SIMP supplied in part V is an example of such a program. There are, however, some difficulties associated with the approach. Since we examine convergence over subintervals of $[a, b]$, we must allocate a permissible error over each subinterval. By this we mean that if we want

$$\int_a^b f(x)\, dx$$

to an accuracy of ϵ, then what accuracy, for example, do we require of approximations to

$$\int_a^{(a+b)/2} f(x)\, dx, \qquad \int_{(a+b)/2}^b f(x)\, dx?$$

A natural choice is an accuracy of $\epsilon/2$ for each. Similarly, if it is necessary to split further, we have so far implicitly allocated the error by

$$\left| \int_\alpha^\beta f(x)\, dx - Q(f) \right| \leq \left| \frac{\beta - \alpha}{b - a} \right| \epsilon.$$

If we denote the approximate integral which finally passes this test by $Q[f, \alpha, \beta]$, then we find

$$\left| \int_a^b f(x)\,dx - \sum_{[\alpha,\beta]} Q[f, \alpha, \beta] \right| = \left| \sum_{[\alpha,\beta]} \left(\int_\alpha^\beta f(x)\,dx - Q[f, \alpha, \beta] \right) \right|$$

$$\leq \sum_{[\alpha,\beta]} \left| \int_\alpha^\beta f(x)\,dx - Q[f, \alpha, \beta] \right| \leq \sum_{[\alpha,\beta]} \left| \frac{\beta - \alpha}{b - a} \right| \epsilon = \epsilon.$$

This is certainly a natural way of allocating an error ϵ among the subintervals, but experience shows that the final sum is usually far more accurate than ϵ, implying that one has been working unnecessarily hard. That this might happen is easily understood. The allocation as stated assumes that the separate errors add up to the largest possible value. We might expect the individual quadratures to err on the high side as often as on the low, with the consequence that the errors partially cancel each other in the final sum. Experience suggests that when splitting an interval in half, the error allowed be split by roughly $1/\sqrt{2}$ instead of $1/2$, for efficiency's sake.

The device for estimating the error that we discussed in the context of iterative methods applies equally well here and is even more important. By accumulating an estimate of the error and returning it to the user, we can compensate for an error allocation procedure based on magnitudes of local errors. In particular, if the device of allocating with a factor of $1/\sqrt{2}$ is unsuitable for a given problem, we may be warned that the accuracy request was not actually met.

Measures of Error

So far we have been speaking of the absolute error,

$$\left| \int_a^b f(x)\,dx - Q(f) \right|,$$

because it is simple and familiar. Yet it is not very suitable for a general purpose code because it depends on the scaling of $f(x)$. More often one is interested in the relative error given by

$$\frac{\left| \int_a^b f(x)\,dx - Q(f) \right|}{\left| \int_a^b f(x)\,dx \right|}.$$

In principle this choice is fine, but it is somewhat unsatisfactory in practice. Suppose that Q_1 and $Q_{11} + Q_{12}$ are approximations obtained by Simpson's rule before and after bisection. The natural approximation to use for the test is then

$$\frac{|(Q_{11} + Q_{12}) - Q_1|}{|Q_{11} + Q_{12}|} \leq \epsilon.$$

The difficulty is that this kind of test can be quite sensitive to computational errors. If the values of $f(x)$ are not small but its integral is, then we might well have $Q_{11}(f) \doteq -Q_{12}(f)$ so that there is severe cancellation in the denominator. The errors in Q_{11} and Q_{12} become relatively more important then and the test is not reliable.

The error test

$$\frac{\left| \int_a^b f(x)\,dx - Q(f) \right|}{\int_a^b |f(x)|\,dx},$$

which is called a relative L_1 error test, avoids the difficulties of a relative error test, and so our code computes an approximation to $\int_a^b |f(x)|\,dx$ as well as $\int_a^b f(x)\,dx$. This costs no additional function evaluations since we need merely form the absolute values of the ones we compute for $f(x)$. Still, this is a bit annoying since the price we pay for a fairly reliable estimate of the error is a rather unfamiliar measure of the error. In further defense of the choice we point out that if $f(x)$ is of one sign only, the L_1 error is the same as relative error. Another important situation is when integrating something like a highly oscillatory function. Then $f(x)$ may not be small but may have an integral that is. This error measure is relative to the integral of the absolute value, which will not be small, with the result that the code will not attempt unrealistic accuracies for these difficult problems.

Some Devices for Integration

We have been discussing the design of a code which basically approximates the integrand by a continuous function made up of pieces of quadratic polynomials and then integrates this function. Many problems arise in a form for which this basic approach is unreasonable, and so we must transform them before our scheme for numerical integration can be used. Here we shall examine a few devices for this transformation. Throughout the numerical examples, "accuracy" means in the relative L_1 sense.

A. Splitting the Interval. If we wish to integrate $f(x)$ over $[a, b]$ and if $f(x)$ is discontinuous at a point c in the interval, the problem is obviously unsuitable for our code. It is easily handled by considering the two problems

$$\int_a^c f(x)\,dx \quad \text{and} \quad \int_c^b f(x)\,dx,$$

using the code to solve each problem separately, and adding the results. If a derivative of $f(x)$ is discontinuous at c but $f(x)$ is not, this kind of behavior can be imitated provided the adaptive code itself splits at c. If we are in doubt whether this will happen or know it cannot, then we must split the problem ourselves. For example, if

$$f(x) = \begin{cases} 1 & \text{if } x \geq \frac{1}{3} \\ -1 & \text{if } x < \frac{1}{3}, \end{cases}$$

the problem

$$\int_0^1 f(x)\,dx$$

will not be integrated exactly by our bisection code (why?), but the two problems

$$\int_0^{1/3} f(x)\,dx, \qquad \int_{1/3}^1 f(x)\,dx$$

will.

B. Change of Variables. A change of variables may be necessary or desirable. If the range of integration is infinite, it is necessary to do something to prepare it for our code. For example, if we want

$$\int_a^\infty f(x)\,dx \tag{9}$$

and $a > 0$, we could let $t = 1/x$ and get the equivalent form

$$\int_0^{1/a} g(t)\,dt$$

where

$$g(t) = \frac{f(1/t)}{t^2}.$$

We have to ask what happens to the special point $x = \infty$. It is mapped into $t = 0$ and

$$\lim_{t\to 0} g(t) = \lim_{x\to\infty} x^2 f(x).$$

Since something like

$$\lim_{x\to\infty} x^{1+\varepsilon} f(x) = 0$$

with an $\epsilon > 0$ is necessary for the integral even to exist, it is often the case that the limit value exists though we may have to be careful in its computer evaluation. As an example, consider the integral

$$\int_1^\infty \frac{e^{-x}}{x}\,dx.$$

Letting $t = \frac{1}{x}$, this transforms into

$$\int_0^1 \frac{e^{-1/t}}{t}\, dt.$$

Here $t = 0$ is an ordinary point for $g(t) = \frac{e^{-1/t}}{t}$ since

$$\lim_{t\to 0} g(t) = 0.$$

This particular transformation is pointless if $a = 0$ in (9), in which case the transformation $t = e^{-x}$ may be useful. Then the transformed integral is

$$\int_0^1 \frac{f(-\log t)\, dt}{t} = \int_0^1 h(t)\, dt.$$

If an integrand becomes infinite at some point, we may be able to remove the singularity with a change of variables. Suppose in

$$\int_a^b f(x)\, dx$$

that $f(x)$ is infinite at $x = c$ for $a \le c \le b$. Split the integral at c first and handle both parts in the same way. Suppose that

$$f(x) = \frac{g(x)}{(x - c)^\alpha}$$

where $g(c)$ exists and is non-zero and that $0 < \alpha < 1$. If we let $(x - c) = t^\beta$ and choose β so that $n = \beta(1 - \alpha)$ is a positive integer, then

$$\int_c^b \frac{g(x)}{(x - c)^\alpha}\, dx = \beta \int_0^{(b-c)^{1/\beta}} g(c + t^\beta)t^{n-1}\, dt. \tag{10}$$

The new integrand is continuous (and so finite) if g is. A special case is when α is the ratio m/k of two integers; then we can take $\beta = k$ and $n = k - m \ge 1$. The new integrand is then as smooth as g itself.

The integral

$$\int_0^1 \frac{e^x}{\sqrt{x}}\, dx \tag{11a}$$

furnishes a simple example, for the change of variables $x = t^2$ yields the new integral

$$2\int_0^1 e^{t^2}\, dt \tag{11b}$$

which is easy to evaluate. The code SIMP with a requested accuracy of 10^{-5} returned the answer 2.925301 for (11b). The estimated accuracy was 6.8×10^{-5}, and 25 function evaluations were required. The exact value correct to seven figures is 2.925303.

C. Subtracting out Singularities. The idea here is to write $f(x) = f_1(x) + f_2(x)$ where $f_1(x)$ is relatively smooth and $f_2(x)$ contains the awkward part of $f(x)$ but can be integrated analytically. Then we evaluate

$$\int_a^b f(x)\,dx = \int_a^b f_1(x)\,dx + \int_a^b f_2(x)\,dx$$

by computing the first integral numerically and the second analytically.

We look first at the simple example $\int_0^1 \sin\sqrt{x}\,dx$. The difficulty with this integrand is that it has an infinite derivative at the origin. The fact that $\sin\sqrt{x} \doteq \sqrt{x}$ for small x suggests subtracting out the "singularity." So we have

$$\int_0^1 \sin\sqrt{x}\,dx = \int_0^1 [\sin\sqrt{x} - \sqrt{x}]\,dx + \int_0^1 \sqrt{x}\,dx. \tag{12}$$

The first integral is evaluated numerically and the second analytically. The new integrand $\sin\sqrt{x} - \sqrt{x}$ is differentiable at $x = 0$ since

$$\frac{d}{dx}(\sin\sqrt{x} - \sqrt{x}) = \frac{\cos\sqrt{x} - 1}{\sqrt{x}}$$

and

$$\lim_{x\to 0} \frac{\cos\sqrt{x} - 1}{\sqrt{x}} = 0.$$

There is a loss of significance in evaluating $\sin\sqrt{x} - \sqrt{x}$ for x near zero since we subtract two nearly equal quantities. This is unimportant, however, since the integrand tends to zero and has little effect on the value of the integral.

Our code will apply directly to the integral $\int_0^1 \sin\sqrt{x}\,dx$. The code SIMP with a requested accuracy of 10^{-5} returned an answer of 0.6023364. The estimated accuracy was 5.0×10^{-7}, and 65 function evaluations were required. The true value of the integral is $2[\sin(1) - \cos(1)] = 0.6023374$; hence the true accuracy is 1.7×10^{-6}. The second integral in (12) was computed to be -0.06432933 with 33 evaluations and an estimated error of 2.0×10^{-8}. This leads to an answer of 0.6023373, an estimated accuracy of 3.0×10^{-8}, and a true accuracy of 1.5×10^{-7}. As a matter of interest, the change of variables $t^2 = x$ leads to the new integral

$$\int_0^1 2t \sin t\,dt.$$

The code returned an answer of 0.6023364 for this integral with an estimated accuracy of 5.2×10^{-7} and a true accuracy of 1.7×10^{-6}; 17 function evaluations were required.

In circumstances of some generality we can see how to get a suitable splitting of $f(x)$. Suppose that

$$f(x) = S(x)g(x),$$

where $S(x)$ behaves badly at c in $[a, b]$ and we can obtain the first few derivatives of $g(x)$ at c. Then we can take

$$f_2(x) = S(x)\left[g(c) + (x - c)g'(c) + \cdots + \frac{(x - c)^n}{n!} g^{(n)}(c)\right],$$

$$f_1(x) = f(x) - f_2(x).$$

To use this procedure we must be able to evaluate the integrals

$$\int_a^b f_2(x)\, dx = \sum_{k=0}^{n} \int_a^b S(x) \frac{(x - c)^k}{k!} g^{(k)}(c)\, dx$$

analytically. Then we compute numerically

$$\int_a^b f_1(x)\, dx = \int_a^b S(x)\left[g(x) - \sum_{k=0}^{n}(x - c)^k \frac{g^{(k)}(c)}{k!}\right] dx.$$

The previous example can also be treated in this way by writing

$$f(x) = \sin \sqrt{x} = \sqrt{x} \cdot \frac{\sin \sqrt{x}}{\sqrt{x}}.$$

If we take $g(x) = \dfrac{\sin \sqrt{x}}{\sqrt{x}}$, $c = 0$ and $n = 0$, we find

$$f_2(x) = \sqrt{x} \cdot 1,$$

$$f_1(x) = \sin \sqrt{x} - \sqrt{x}$$

which is a splitting already treated. If we take $n = 1$, we find

$$f_2(x) = \sqrt{x}\left(1 - \frac{x}{6}\right),$$

$$f_1(x) = \sin \sqrt{x} - f_2(x).$$

With a requested accuracy of 10^{-5}, the integral of $f_1(x)$ was found to be 0.002337339. The estimated error was 1.1×10^{-7} and 25 evaluations were required. This leads to an answer of 0.6023374 with an estimated accuracy of 1.8×10^{-9}.

Example. Let us now discuss a realistic example in detail. A conducting ellipsoidal column projecting above a flat conducting plane has been used as a model of a lightning rod [4] (Fig. II.2). When the ellipsoid is given by the equation

$$\frac{x^2}{a^2} + \frac{y^2}{b^2} + \frac{z^2}{c^2} = 1,$$

the potential function is

$$V(x, y, z) = -z + Az \int_\lambda^\infty \frac{du}{\sqrt{(a^2 + u)(b^2 + u)(c^2 + u)^3}}.$$

The constant A depends only on the shape of the rod and is given by

$$A = 1 \bigg/ \int_0^\infty \frac{du}{\sqrt{(a^2 + u)(b^2 + u)(c^2 + u)^3}}.$$

The quantity λ is a function of the place where we want to evaluate the potential. It is the unique non-negative root of the equation

$$\varphi(\lambda) = \frac{x^2}{a^2 + \lambda} + \frac{y^2}{b^2 + \lambda} + \frac{z^2}{c^2 + \lambda} - 1 = 0.$$

Suppose that for a tall, slender rod described by $a = 1$, $b = 2$, $c = 100$ we seek the potential V at $x = 50$, $y = 50$, $z = 50$. As we shall see in the next chapter, we compute λ to be approximately 5928.359. To compute

$$\int_\lambda^\infty \frac{du}{\sqrt{(a^2 + u)(b^2 + u)(c^2 + u)^3}}, \tag{13}$$

we note that the integrand tends to zero like $u^{-5/2}$ as $u \to \infty$ so the integral is well defined and the change of variables $t = 1/u$ is satisfactory. A straightforward substitution leads to

$$\int_0^{1/\lambda} \frac{dt}{t^2\sqrt{(a^2 + 1/t)(b^2 + 1/t)(c^2 + 1/t)^3}}.$$

This is perfectly satisfactory mathematically, but is in a bad computational form. Here it is easy to rearrange to obtain

$$\int_0^{1/\lambda} \sqrt{\frac{t}{(a^2t + 1)(b^2t + 1)(c^2t + 1)^3}}\, dt.$$

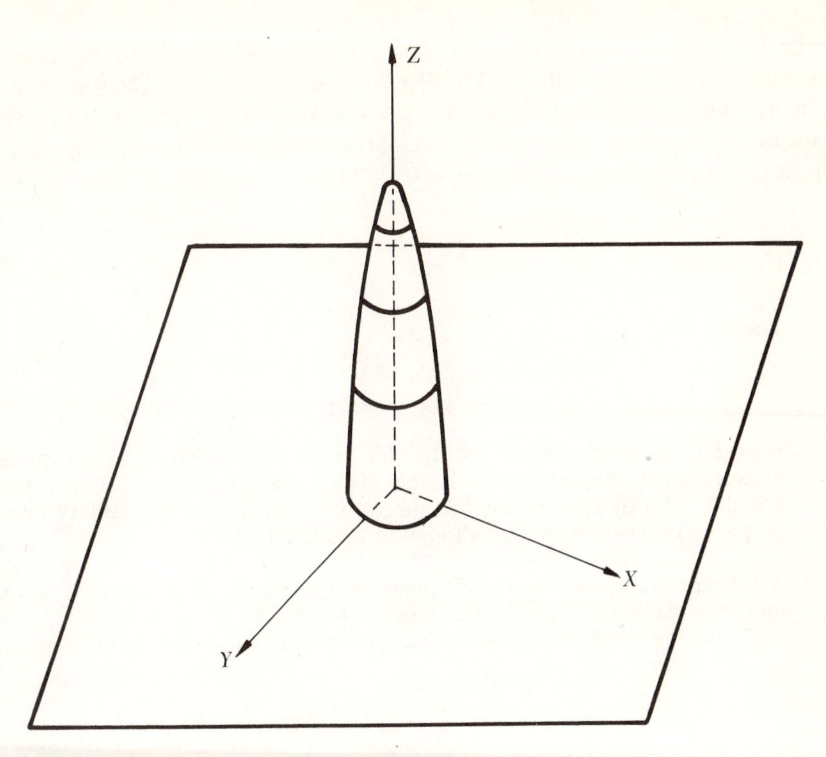

Figure II.2

Using the code SIMP and a requested accuracy of 10^{-5}, we obtained 5.705946×10^{-7} with an estimated accuracy of 6.9×10^{-7} at a cost of 77 function evaluations. The function is approximately $\sqrt{t}$ near $t = 0$, which is a little difficult because of the infinite derivative at zero. This suggests we have not made the best change of variables, and perhaps we should also use $w^2 = t = 1/u$. Then the integral is

$$\int_0^{1/\sqrt{\lambda}} \frac{2w^2\, dw}{\sqrt{(a^2w^2 + 1)(b^2w^2 + 1)(c^2w^2 + 1)^3}}.$$

This is a somewhat easier integral to evaluate. We found a value of 5.705914×10^{-7} with an estimated accuracy of 1.8×10^{-7} at a cost of 33 function evaluations. Note that the two results are consistent to about five figures.

These transformations cannot be applied directly to computing A because the interval remains infinite (work it out!). Since we have to compute (13) anyway, we can split

$$\int_0^{\infty} \frac{du}{\sqrt{(a^2 + u)(b^2 + u)(c^2 + u)^3}} = \int_0^{\lambda} + \int_{\lambda}^{\infty}.$$

The first integral is computed to be 7.218085×10^{-6} with an estimated accuracy of 8.8×10^{-7} at a cost of 213 function evaluations. The sum using the second value computed above is 7.788616×10^{-6} with an estimated accuracy of 8.3×10^{-7}, which gives $A = 1.283925 \times 10^{5}$. Using these results, we find the potential $V = -46.3370$.

EXERCISES

1. The function

$$y(x) = e^{-x^2} \int_0^x e^{t^2}\, dt$$

is called Dawson's integral. In II.4 this function is evaluated by solving a differential equation. Make up a table of this function for $x = 0.0, 0.1, 0.2, 0.3, 0.4, 0.5$. By splitting up the integral you will not do unnecessary integrations. Compare with results found in standard tables [1].

2. A sphere of radius R floats half submerged in a liquid. If it is pushed down until the diametral plane is a distance p (such that $0 < p \leq R$) below the surface of the liquid and is then released, the period of the resulting vibration is

$$T = 8R \sqrt{\frac{R}{g(6R^2 - p^2)}} \int_0^{2\pi} \frac{dt}{\sqrt{1 - k^2 \sin^2 t}},$$

$$k^2 = \frac{p^2}{6R^2 - p^2} \cdot$$

where $g = 32.174$ ft./sec.^{2}. For $R = 1$, find T when $p = 0.50, 0.75, 1.00$.

3. Write a code to use the compound Simpson rule in the iterative manner suggested. Compare its efficiency in terms of the number of function evaluations with that of the iterative, adaptive code SIMP. The variable KOUNT in SIMP counts function evaluations. Use a number of integrals for which you have analytical expressions for the answers.

4. What is the minimum number of function evaluations the code SIMP can make? The problem

$$\int_0^4 x^2(x-1)^2(x-2)^2(x-3)^2(x-4)^2\, dx$$

causes the code to "fail"; explain why. Try it.

5. The procedures of this chapter are aimed at computing $\int_a^b f(x)\, dx$ when $f(x)$ can be evaluated anywhere in $[a, b]$. It is a common problem to compute such an integral when $f(x)$ is known only at points $x_0 < x_1 < \cdots < x_n$; for example these might be the places where an experimenter measures f, and the values $f(x_i)$ are are subject to errors. If the errors are not significant, the procedure using splines developed in II.1.3 problem 2 is very useful. If the errors are significant, we could fit a least squares polynomial as in II.1.2 and integrate

it. An alternative which can also be used when the errors are significant integrates interpolating quadratic polynomials; to smooth out the errors of measurement, $f(x)$ is approximated as the average of two interpolating quadratics. Let $Q[x, x_{i-1}, x_i, x_{i+1}]$ denote the quadratic interpolating to $f(x)$ at x_{i-1}, x_i, x_{i+1}. Then one approximates

$$\int_{x_i}^{x_{i+1}} f(x)\, dx \doteq \tfrac{1}{2} \int_{x_i}^{x_{i+1}} \{Q[x, x_{i-1}, x_i, x_{i+1}] + Q[x, x_i, x_{i+1}, x_{i+2}]\}\, dx.$$

We suppose that $a < x_1$, $x_{n-1} < b$ so that we cannot average on the ends; we just use

$$\int_a^{x_1} f(x)\, dx \doteq \int_a^{x_1} Q[x, x_0, x_1, x_2]\, dx,$$

$$\int_{x_{n-1}}^{b} f(x)\, dx \doteq \int_{x_{n-1}}^{b} Q[x, x_{n-2}, x_{n-1}, x_n]\, dx.$$

Work out all these formulas and implement them in a code. The reference [13] contains further discussion of the idea (pp. 22–24) and a FORTRAN code AVINT (p. 193).

6. Verify that the change of variables in equation (10) for the case $\alpha = m/k$, where m and k are integers and $0 < \alpha < 1$, leads to a new integrand which has a continuous derivative if g does.

7. The exponential integral

$$E_1(t) = \int_1^{\infty} e^{-tx} \frac{dx}{x}, \qquad t > 0,$$

arises in the study of radiative transfer and transport theory [10]. Some manipulation shows that

$$E_1(t) = \int_1^{\infty} e^{-x} \frac{dx}{x} + \int_t^1 e^{-x} \frac{dx}{x}$$

$$= -\left\{\int_1^{\infty} e^{-x} \frac{dx}{x} - \int_0^1 (1 - e^{-x}) \frac{dx}{x}\right\} + \int_t^1 \frac{dx}{x} + \int_0^t (1 - e^{-x}) \frac{dx}{x}.$$

The expression in braces is known to have the value $\gamma \doteq .5772157$, the Euler-Mascheroni constant, and the second term integrates analytically to $(-\ln t)$. So

$$E_1(t) = -\gamma - \ln t + \int_0^t (1 - e^{-x}) \frac{dx}{x}.$$

Using SIMP, evaluate $E_1(t)$ for $t = 1.0, 2.0, 3.0$. Compare with results found in standard tables [1].

8. The potential inside the unit circle due to a specified potential $f(\theta)$ on the boundary is given by Poisson's integral

$$\varphi(r, \theta) = \frac{1}{2\pi} \int_0^{2\pi} \frac{1 - r^2}{1 - 2r\cos(\theta - \theta') + r^2} f(\theta')\, d\theta'.$$

There is difficulty evaluating this integral as $r \to 1$, since for $\theta' = \theta$,

$$\frac{1 - r^2}{1 - 2r\cos(\theta - \theta') + r^2} = \frac{1 + r}{1 - r}.$$

This is not too severe because the term is large only if r is very close to 1, but there should be no difficulty at all since as $r \to 1$, $\varphi(r, \theta) \to f(\theta)$ (see Bateman [4], pp. 239–241). Realizing that

$$1 = \frac{1}{2\pi} \int_0^{2\pi} \frac{1 - r^2}{1 - 2r\cos(\theta - \theta') + r^2}\, d\theta',$$

derive the form

$$\varphi(r, \theta) = f(\theta) + \frac{1}{2\pi} \int_0^{2\pi} \frac{1 - r^2}{1 - 2r\cos(\theta - \theta') + r^2} [f(\theta') - f(\theta)]\, d\theta'$$

and argue that it should have somewhat better numerical properties. Explore this by computing for r approaching 1 with $f(\theta) = \sin\theta$. The analytical solution is then just $\varphi(r, \theta) = r\sin\theta$.

9. The potential in a conducting strip of width b with potential zero on the bottom edge and a specified potential $F(x)$ on the upper edge is

$$\varphi(x, y) = \frac{1}{b}\sin(\pi y/b) \int_{-\infty}^{\infty} \frac{F(\xi)\, d\xi}{\cosh[(\xi - x)\pi/b] + \cos(\pi y/b)}.$$

Suppose that an experimenter applies the potential $F(x) = 1$ for $|x| \leq 0.99$ and $F(x) = \exp[-100(|x| - 0.99)]$ for $|x| \geq 0.99$. When $b = \pi$, compute and plot the potential along the middle of the strip, $\varphi\left(x, \frac{\pi}{2}\right)$.

Realizing that

$$\cosh[(\xi - x)\pi/b] \geq 1,$$

bound the effect on $\varphi(x, y)$ for $y \neq 0$ of replacing the infinite interval by a finite one

$$\int_{-z}^{z} \frac{F(\xi)\, d\xi}{\cosh[(\xi - x)\pi/b] + \cos(\pi y/b)}.$$

For a suitable choice of z, use this instead of the infinite interval. Show for the given $F(x)$ that $\varphi(x, y) = \varphi(-x, y)$, so only a plot for $x \geq 0$ is necessary.

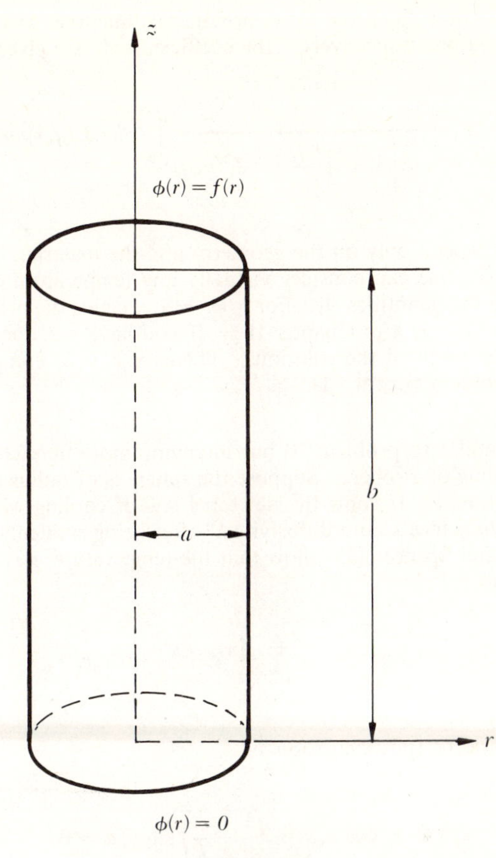

Figure II.3

10. This example is representative of a great many computations arising in the use of the classical separation of variables technique for solving field problems. Typically, one must compute many roots of nonlinear equations and integrals. A detailed discussion of this particular problem may be found in Moon and Spencer, *Field Theory for Engineers* [33]. The temperature distribution in a cylinder (Fig. II.3) of radius a and height b with the bottom held at a temperature zero, the top at a temperature $f(r)$, and the side dissipating heat according to Newton's law of cooling, can be represented by a series. If the thermal conductivity of the cylinder is k and the thalpance is ϵ, then the temperature $\varphi(r, z)$ is

$$\varphi(r, z) = \sum_{n=1}^{\infty} A_n \frac{\sinh(q_n z)}{\sinh(q_n b)} J_0(q_n r).$$

The numbers q_n are the positive roots of the equation

$$\frac{k}{\epsilon a} q_n a J_1(q_n a) - J_0(q_n a) = 0$$

where the function $J_0(x)$ and $J_1(x)$ are Bessel functions of the first kind of orders zero and one respectively. The coefficients A_n are given by

$$A_n = \frac{2}{a^2\left[1 + \left(\frac{kq_n}{\epsilon}\right)^2\right] J_1^2(q_n a)} \int_0^a r f(r) J_0(q_n r)\, dr.$$

The roots q_n depend only on the geometry and the material. Once they have been computed, one can consider virtually any temperature distribution $f(r)$ by computing the quantities A_n. For $k/\epsilon a = 2$, we give the problem of solving for $q_n a$ for $n = 1, 2, 3$ in Chapter II.3. If you have not done that problem, use the results given in the solutions. Then for $a = 1$, compute A_1, A_2, A_3 for $f(r) = \exp(-r) - \exp(-1)$.

11. A problem similar to problem 10 but involving more familiar functions concerns the cooling of a sphere. Suppose the sphere is of radius a and is initially at a temperature V. It cools by Newton's law of cooling with thermal conductivity k, thalpance ϵ, and diffusivity h^2 after being suddenly placed in air at 0°C. Moon and Spencer [33] show that the temperature $\varphi(r, t)$ at time $t > 0$ and at radius r is

$$\varphi(r, t) = \sum_{n=1}^{\infty} \frac{A_n}{r} e^{-\gamma_n^2 h^2 t} \sin \gamma_n r.$$

Here the γ_n are the (positive) roots of

$$\gamma_n \cos \gamma_n a - \left(\frac{1}{a} - \frac{\epsilon}{k}\right) \sin \gamma_n a = 0$$

and

$$A_n = \frac{2\gamma_n V}{[\gamma_n a - \cos \gamma_n a \sin \gamma_n a]} \int_0^a r \sin \gamma_n r\, dr.$$

For a steel sphere cooling in air at 0°C, suppose the initial temperature is $V = 100$°C and the radius is $a = 0.30$ meters. Appropriate physical constants are $h^2 = 1.73 \times 10^{-5}$, $\epsilon = 20$, $k = 60$. Then $\epsilon a/k = 0.10$ and the first few values of $\gamma_n a$ are found to be

n	$\gamma_n a$
1	.5422819
2	4.515659
3	7.738193

Using these values, compute A_1, A_2, A_3 and evaluate the temperature at $r = 0.25$ for $t = 10^k$ seconds for $k = 2, 3, 4, 5$. One can perform the integration required for A_n analytically. Do this and compare the analytical to the computed values.

12. Use the method of subtracting out the singularity and the code SIMP to evaluate

$$\int_0^1 \frac{e^x}{\sqrt{x}}\,dx.$$

Compare your result with that obtained in the text using a change of variables.

13. In 1829, Dulong and Petit reported that the molar specific heat of metallic solids, C_v, was a constant for many metals and was approximately equal to 6 calories/mole at room temperature (approximately 300°K). This proved very useful to chemists during the following years in assigning unambiguous atomic weights to the elements. However, carbon in the form of diamond was soon found to be an exception to the rule, having an anomalously low specific heat. This had the consequence of discrediting the use of the law of Dulong and Petit when assigning atomic weights, until the anomaly was explained by Einstein in 1905 using the quantum theory. Einstein showed that C_v is in fact a function of temperature T; specifically,

$$C_v(T) = \frac{3R(\theta_E/T)^2 \exp(\theta_E/T)}{(\exp(\theta_E/T) - 1)^2}$$

where $R = 1.986$ calories/mole is called the molar gas constant and θ_E is a parameter characteristic of the substance.

Einstein's specific heat law was improved slightly by Debye in 1913 to give better low temperature behavior. Debye's specific heat law is now used extensively, but unfortunately it is a more complicated function of temperature:

$$C_v(T) = \frac{9R}{U_D^3}\int_0^{U_D} \frac{x^4 e^x}{(e^x - 1)^2}\,dx$$

where $U_D = \theta_D/T$ and θ_D is the Debye temperature characteristic of each substance.

Evaluate the Debye specific heat for diamond at the temperatures indicated below and compare with the experimental values given.

diamond ($\theta_D = 1900$°K)

T (°K)	C_v (cal./mole).
100	0.06
200	0.56
300	1.46
400	2.42
500	3.17
1000	5.09

You may also wish to evaluate specific heat from Einstein's theory using the empirical rule $\theta_E \doteq \frac{3}{4}\theta_D$.

14. Verify that Simpson's rule integrates a cubic polynomial exactly. Hint: Express the cubic polynomial $P(x)$ in terms of its Taylor series expansion about the point $c = \dfrac{b + a}{2}$.

15. In performing an arginine tolerance test, a doctor measures glucose, insulin, glucagon, and growth hormone levels in the blood over a one hour time period at ten minute intervals to obtain the following data:

time	glucose	insulin	glucagon	growth hormone
0	102	11	188	1.70
10	114	26	1300	1.70
20	122	36	2300	1.20
30	132	47	2600	2.50
40	115	39	1800	7.25
50	107	27	840	8.10
60	100	15	460	8.00

The doctor is interested in the integrated effect of each response. For example, if we represent the glucose curve by $g(t)$, he wants

$$\int_0^{60} g(t)\,dt$$

for glucose. Use the method outlined in problem 5 to evaluate the required integrals.

3. ROOTS OF NONLINEAR EQUATIONS

The Problem

We shall discuss the solution of the equation

$$f(x) = 0 \tag{1}$$

when f is a continuous real function of a single real variable x. This is a frequently occurring problem which lends itself nicely to graphical interpretation and which can be discussed in elementary terms. More general problems involving several variables are also important, but it is beyond the scope of this text to discuss them; we refer the interested reader to the references [36, 37, 42].

By a root of (1), or a zero of $f(x)$, we mean a number α such that $f(\alpha) = 0$. A root is described more usefully if we say it is of multiplicity m. By this we mean that for x near α, $f(x)$ can be written in the form

$$f(x) = (x - \alpha)^m g(x) \tag{2}$$

where $g(x)$ is continuous near α and $g(\alpha) \neq 0$. If $m = 1$, we say that the root is simple; otherwise, it is multiple. This basic definition permits m to be a fraction; for example, with the function

$$f(x) = x\sqrt{x - 1},$$

equation (1) has $\alpha = 1$ as a root of multiplicity 1/2 (and $\alpha = 0$ as a simple root). However, if $f(x)$ has sufficiently many continuous derivatives near α, then m must be a positive integer. In fact, if $f(x)$ has its first k derivatives continuous at α and

$$\begin{cases} f(\alpha) = 0, \\ f'(\alpha) = f''(\alpha) = \cdots = f^{(k-1)}(\alpha) = 0, \\ f^{(k)}(\alpha) \neq 0, \end{cases} \tag{3}$$

then α is a root of multiplicity k. To see this, we expand $f(x)$ in a Taylor series about α to obtain

$$f(x) = f(\alpha) + (x - \alpha)f'(\alpha) + \frac{(x - \alpha)^2}{2} f''(\alpha) + \cdots + \frac{(x - \alpha)^{k-1}}{(k - 1)!} f^{(k-1)}(\alpha) + \frac{(x - \alpha)^k}{k!} f^{(k)}(\xi_x),$$

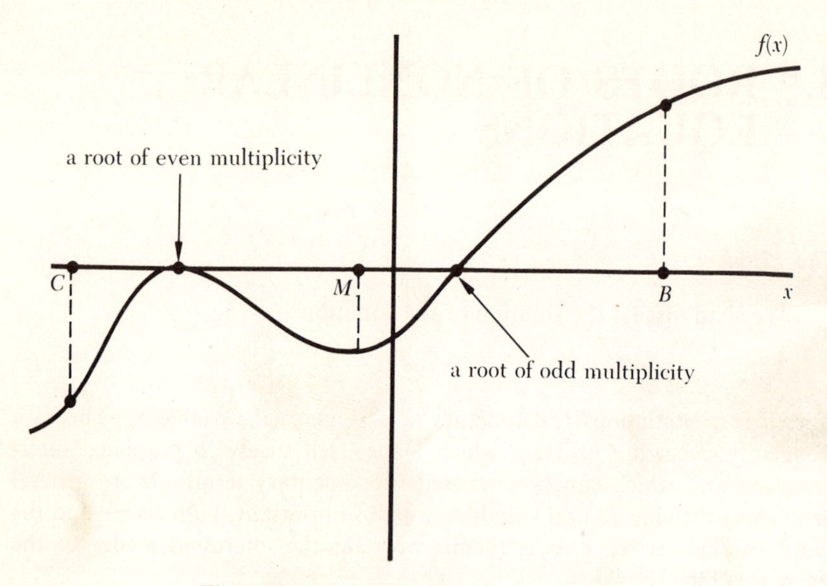

Figure II.4. Graphical interpretation of roots.

where ξ_x lies between x and α. Using (3), this simplifies to

$$f(x) = \frac{(x - \alpha)^k}{k!} f^{(k)}(\xi_x). \tag{4}$$

If we take $g(x) = \dfrac{f^{(k)}(\xi_x)}{k!}$, then $g(\alpha) = \dfrac{f^{(k)}(\alpha)}{k!} \neq 0$. We shall always suppose that $f(x)$ is sufficiently smooth near α that we can use (3) instead of the basic definition (2) and, in particular, we shall always have roots of integer multiplicity.

From the definition of a root α, we see that the graph of $f(x)$ must touch the x axis at α (Fig. II.4). For a root of multiplicity k, the function $f^{(k)}(x)$ does not change sign near α because it is continuous and $f^{(k)}(\alpha) \neq 0$. This observation and the expression (4) shows that if k is even, $f(x)$ is tangent to the x axis at α but does not cross there; if k is odd, $f(x)$ crosses the axis at α.

Bisection, Newton's Method, and the Secant Rule

If we have two points $x = B$ and C where $f(x)$ has opposite signs, the continuity of f implies that it must have at least one zero between them. From the graphical interpretation we can say somewhat more. Zeros of even multiplicity between B and C do not cause a sign change, and zeros of odd multiplicity do. If there were an even number of zeros of odd multiplicity between B and C, the sign changes would cancel out and f would have the same sign at both ends. Thus, if $f(B)f(C) < 0$, there must be an odd number of zeros of odd multiplicity and possibly some zeros of even multiplicity between B and C. If we agree to count the number of zeros according to their multiplicity (i.e., a zero of multiplicity m counts as m distinct zeros), then we see there are an odd number of zeros between B and C.

The method of bisection is based on the use of sign changes to detect a zero. If $f(B)f(C) < 0$, we evaluate $f(x)$ at the midpoint $M = \dfrac{B + C}{2}$. If $f(M) = 0$, we have a zero. Otherwise, either $f(B)f(M) < 0$ or $f(M)f(C) < 0$. In the first case there is at least one zero in $[M, B]$, as in Figure II.4, and in the second case, there is at least one zero in $[C, M]$. This produces an interval with half the length of the original interval in which a zero must lie. We repeat the procedure until we arrive at an interval of sufficiently small length which contains a zero.

There are a number of virtues of this simple procedure. It always converges, at least in the sense that it produces an interval of specified length such that the *computed* function values have opposite signs at the ends of the interval. We emphasize that we refer to the computed function value, since we often try to make the computed function value as small as possible. In attempting this, the finite word length of our computer enters into the process, as well as do the details of the procedure for evaluating f. As we approach the smallest function values that can be obtained with our machine's precision, even the sign of the computed value may well be incorrect; we say then that we are at limiting precision. This point will be examined again later. It is an important virtue of bisection that errors of this sort do not cause it to give absurd approximations, so we call it stable with respect to limiting precision. The rate of convergence is unaffected by the multiplicity of the zero we are computing, in contrast to other methods we shall discuss. The bracketing of a zero in an interval is very important, since it allows us to decide easily and reliably when convergence has taken place.

Bisection has several serious disadvantages. If there are an even number of zeros between B and C, it will not realize that there are any zeros at all because there is no sign change. Of course, if one begins bisecting, he may detect a sign change eventually and calculate a zero as usual. But it is clear that it is not possible to find a zero of even multiplicity in this way except by accident. The other major disadvantage is that for simple zeros, which seem to be the most common by far, there are methods which converge much more rapidly. Let us now consider two methods which are superior to bisection in these respects (they have other problems, though!). Both approximate $f(x)$ by a straight line $L(x)$ and hence approximate a root of $f(x) = 0$ by a root of $L(x) = 0$.

Newton's method (Fig. II.5) may be familiar to the reader from his study of calculus. It takes $L(x)$ as the tangent line to $f(x)$ at the latest approximation x_i, and the next approximation (iterate) is the root x_{i+1} of $L(x) = 0$. Equivalently, approximating $f(x)$ by the linear terms of a Taylor's

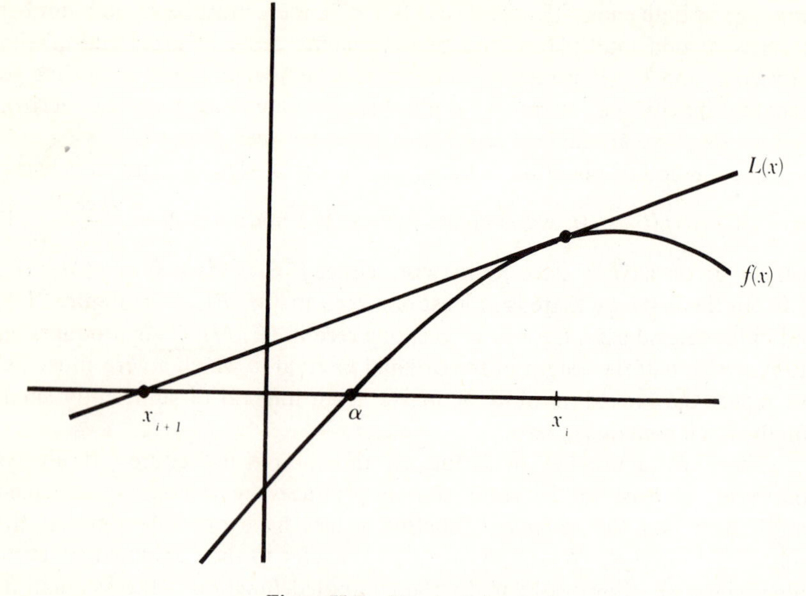

Figure II.5. Newton's method.

series about x_i,

$$f(x) \doteq f(x_i) + (x - x_i)f'(x_i),$$

suggests solving

$$f(x_i) + (x - x_i)f'(x_i) = 0$$

for its root x_{i+1} to approximate α:

$$x_{i+1} = x_i - \frac{f(x_i)}{f'(x_i)}. \tag{5}$$

Often, either we cannot evaluate $f'(x)$ or it is very expensive, so we consider a procedure similar to Newton's method which uses only values of $f(x)$. Suppose we have two approximations x_{i-1}, x_i; the secant of $f(x)$ is the line interpolating $f(x)$ at these points:

$$L(x) = f(x_i) + (x - x_i)\frac{f(x_i) - f(x_{i-1})}{x_i - x_{i-1}}.$$

The next approximation x_{i+1} is taken to be the root of $L(x) = 0$, that is,

$$x_{i+1} = x_i - f(x_i)\frac{x_i - x_{i-1}}{f(x_i) - f(x_{i-1})}. \tag{6}$$

This method, called the secant rule, is illustrated graphically in Figure II.6. Although a picture furnishes a natural motivation for this method, an

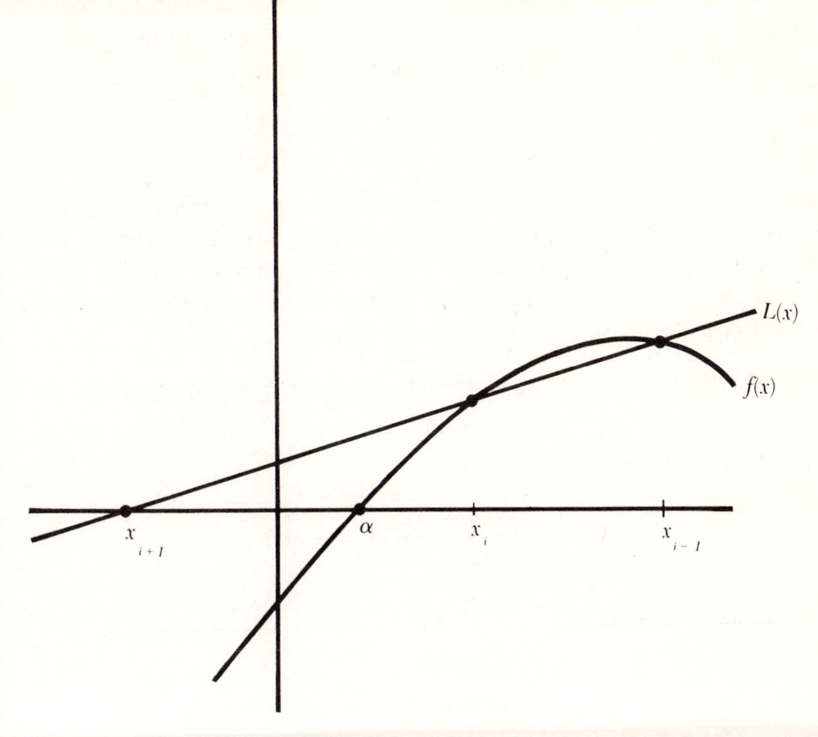

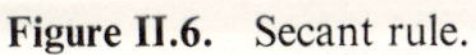
Figure II.6. Secant rule.

alternative approach to it is just to approximate the derivative in Newton's method (5) by a divided difference to get (6).

We want now to indicate why Newton's method and the secant rule converge faster than bisection for a simple root of (1). Considering first Newton's method, we have from (5)

$$x_{i+1} - \alpha = x_i - \alpha - \frac{f(x_i)}{f'(x_i)}.$$

If x_i is near α, then

$$f(x_i) \doteq f(\alpha) + (x_i - \alpha)f'(\alpha) + \frac{(x_i - \alpha)^2}{2} f''(\alpha),$$

$$f'(x_i) \doteq f'(\alpha).$$

Since $f(\alpha) = 0$, and $f'(\alpha) \neq 0$,

$$x_{i+1} - \alpha \doteq x_i - \alpha - \frac{(x_i - \alpha)f'(\alpha) + \dfrac{(x_i - \alpha)^2}{2} f''(\alpha)}{f'(\alpha)}$$

$$= -(x_i - \alpha)^2 \frac{f''(\alpha)}{2f'(\alpha)}.$$

Thus, if x_i is near a simple root, the error in x_{i+1} is roughly a constant multiple of the square of the error in x_i. We call this type of convergence quadratic.

A similar look at the secant rule (6) leads to

$$x_{i+1} - \alpha \doteq (x_i - \alpha)(x_{i-1} - \alpha)\frac{f''(\alpha)}{2f'(\alpha)}. \tag{7}$$

Since x_i should have a smaller error than x_{i-1}, the convergence here is not as fast as for Newton's method, but it is faster than for bisection. With both methods one can show that if the starting values are sufficiently close to a simple root and $f(x)$ is sufficiently smooth, the iterates (the quantities x_i) will converge to the root. For Newton's method,

$$\lim_{x_i \to \alpha} \frac{(x_{i+1} - \alpha)}{(x_i - \alpha)^2} = C \neq 0$$

and for the secant rule

$$\lim_{x_i \to \alpha} \frac{(x_{i+1} - \alpha)}{(x_i - \alpha)(x_{i-1} - \alpha)} = c \neq 0.$$

A careful treatment of the secant rule even shows that

$$\lim_{x_i \to \alpha} \frac{|x_{i+1} - \alpha|}{|x_i - \alpha|^p} = \gamma \neq 0$$

where $p = \dfrac{1 + \sqrt{5}}{2} \doteq 1.618$. For the iterates of bisection, it is clear that

$$\left| \frac{x_{i+1} - \alpha}{x_i - \alpha} \right| \leq \tfrac{1}{2}$$

with equality possible at every step. When the limiting behavior is evident, the secant rule is considerably faster than bisection, and Newton's method is faster still.

**Convergence of the Secant Rule*

The advanced text by Householder [24] gives a convergence result for the secant rule which is both illuminating and elementary. The proof shows that in suitable circumstances $f(x_i) \to 0$. As a first step in the proof we derive an expression which relates the function values at three successive iterates x_{i-1}, x_i, x_{i+1}. As you will recall, we have already used $L(x)$ to denote the polynomial of degree one interpolating to $f(x)$ at x_i and x_{i-1}. We defined the iterate x_{i+1} to be the zero of $L(x)$. In Chapter II.1.1 we developed an expression for the error in interpolation, which in this case is

$$f(x_{i+1}) - L(x_{i+1}) = (x_{i+1} - x_{i-1})(x_{i+1} - x_i)\frac{f''(z)}{2}$$

or, since $L(x_{i+1}) = 0$,

$$f(x_{i+1}) = (x_{i+1} - x_{i-1})(x_{i+1} - x_i)\frac{f''(z)}{2} \tag{8}$$

for a suitable (unknown) point z. Some manipulation using equation (6) gives the two relations

$$x_{i+1} - x_i = \frac{(x_{i-1} - x_i)f(x_i)}{f(x_i) - f(x_{i-1})}, \tag{9a}$$

$$x_{i+1} - x_{i-1} = \frac{(x_{i-1} - x_i)f(x_{i-1})}{f(x_i) - f(x_{i-1})}. \tag{9b}$$

A third relation is obtained from the mean value theorem for derivatives:

$$\frac{f(x_i) - f(x_{i-1})}{x_i - x_{i-1}} = f'(\xi) \tag{9c}$$

where ξ, a point between x_i and x_{i-1}, is unknown. Combining equations (8) and (9), we arrive at

$$f(x_{i+1}) = f(x_i)f(x_{i-1})\frac{f''(z)}{2[f'(\xi)]^2}.$$

A convergence proof is going to require bounds saying that $|f(x_{i+1})|$ is "smaller" than $|f(x_i)|$ and $|f(x_{i-1})|$. Let us assume that on some appropriate interval

$$|f''(x)| \le M_2,$$

$$0 < m_1 \le |f'(x)| \le m_2,$$

and that we are computing a simple zero α. (Why *must* it be simple with these hypotheses?) Then these bounds and the expression above for $f(x_{i+1})$ imply

$$|f(x_{i+1})| \le |f(x_i)|\,|f(x_{i-1})|\frac{M_2}{2m_1^2}.$$

If we let

$$\epsilon_i = |f(x_i)|\frac{M_2}{2m_1^2},$$

this inequality leads to

$$\epsilon_{i+1} \le \epsilon_i\epsilon_{i-1}.$$

Suppose

$$\epsilon = \max(\epsilon_0, \epsilon_1) < 1,$$

that is, the initial "error" is small. Then it is easy to argue by induction that

$$\epsilon_2 \leq \epsilon^2$$
$$\epsilon_3 \leq \epsilon^2 \cdot \epsilon = \epsilon^3$$
$$\epsilon_4 \leq \epsilon^3 \cdot \epsilon^2 = \epsilon^5$$
$$\vdots$$
$$\epsilon_i \leq \epsilon^{\delta_i}$$

where

$$\delta_i = \frac{1}{\sqrt{5}}\left[\left(\frac{1+\sqrt{5}}{2}\right)^{i+1} - \left(\frac{1-\sqrt{5}}{2}\right)^{i+1}\right].$$

We leave the proof of this as an exercise. Since

$$\left|\frac{1+\sqrt{5}}{2}\right| > 1 > \left|\frac{1-\sqrt{5}}{2}\right|,$$

we see that for i large, $\delta_i \doteq \frac{1}{\sqrt{5}}\left(\frac{1+\sqrt{5}}{2}\right)^{i+1}$. In any event, $\delta_i \to \infty$, and since $0 \leq \epsilon < 1$, we must have $\epsilon_i \to 0$, which is what we wanted to prove. Let us now state a formal theorem and complete the details of its proof.

Theorem. The secant rule defined by (6) with initial guesses x_0, x_1 converges to a simple zero α of $f(x)$ if x_0, x_1 lie in a sufficiently small interval containing α on which $f'(x), f''(x)$ exist and do not vanish.

PROOF. The result follows easily from the preceding arguments. First observe that with the assumptions on f' and f'' the bounds m_1, m_2 and M_2 are well defined. Using the mean value theorem for derivatives, we see that

$$|f(x_0)| = |f(\alpha) + (x_0 - \alpha)f'(\zeta_1)| \leq |x_0 - \alpha|\, m_2,$$
$$|f(x_1)| = |f(\alpha) + (x_1 - \alpha)f'(\zeta_2)| \leq |x_1 - \alpha|\, m_2.$$

This implies that the quantity ϵ defined above is less than 1 if x_0 and x_1 are sufficiently close to α. The proof above shows that

$$|f(x_i)|\, \frac{M_2}{2m_1^2} \leq \epsilon^{\delta_i}.$$

But

$$|f(x_i)| = |f(x_i) - f(\alpha)| = |(x_i - \alpha)f'(\eta)|$$
$$\geq |x_i - \alpha|\, m_1$$

Hence,

$$|x_i - \alpha| \leq \frac{2m_1}{M_2} \epsilon^{\delta_i}.$$

This says that $x_i \to \alpha$. The proof suggests that the order of convergence is $\frac{1}{2}(1 + \sqrt{5})$, as we stated earlier.

Practical Problems of Newton's Method and the Secant Rule

Newton's method and the secant rule have some disadvantages compared to bisection. They do not have the bracketing property, and convergence is questionable. For multiple roots their convergence is no faster than bisection. There are some special situations that one must beware of. Either method can lead to extremely large iterates because the tangent or secant line can become nearly horizontal. As an excellent example of this, consider the problem

$$x^{20} - 1 = 0.$$

In attempting to compute the simple root $\alpha = 1$ using Newton's method, suppose we have $x_0 = 1/2$; then from (5)

$$x_1 = \tfrac{1}{2} - \frac{(\frac{1}{2})^{20} - 1}{20(\frac{1}{2})^{19}} = 26213.875$$

because the tangent is nearly horizontal. Thus, a reasonably good guess leads to a *much* worse approximation. Also, notice that if x_i is much larger than 1, then

$$x_{i+1} = x_i - \frac{x_i^{20} - 1}{20x_i^{19}} \doteq x_i - \frac{x_i^{20}}{20x_i^{19}} = \tfrac{19}{20}x_i.$$

To the same degree of approximation

$$\frac{x_{i+1} - 1}{x_i - 1} \doteq \frac{x_{i+1}}{x_i} \doteq \tfrac{19}{20},$$

which says that we creep back to the root at 1 at a rate considerably *slower* than bisection. This example points out both the problem of horizontal tangents and the fact that one can assert fast convergence only when he is near a root.

A similar phenomenon occurs when we are near the limits of precision of our particular computer. Typically, there is an interval of machine representable numbers about α which lead to computed values of the function which vary erratically in sign and magnitude. These represent the smallest values the computed $f(x)$ can assume using the given precision, and quite frequently they have no digits in agreement with the true $f(x)$. For a simple root, $f'(\alpha) \neq 0$, and so one usually gets a few correct digits in the computed value of the derivative in this interval. As a consequence, Newton's method yields a sequence of iterates which stay in this interval of limiting precision. On the other hand, we see from (6) that the secant rule calculates

$$\frac{x_i - x_{i-1}}{f(x_i) - f(x_{i-1})},$$

which could give unpredictable values at limiting precision. In particular, an iterate may be carried far outside the interval of limiting precision. (We remark in passing that this ratio is subject to severe cancellation, but away from limiting precision the correspondingly small value of $f(x_i)$ makes the loss of significance unimportant in computing the next iterate.) The same thing happens to Newton's method when computing multiple roots.

An Algorithm Combining Bisection and the Secant Rule

It is a challenging task to fuse these methods into an efficient computational scheme. We supply a code, ZEROIN, based upon a code written by Dekker [14]. Our modifications of his excellent code are small but useful. The code is based upon bisection and the secant rule so that only the ability to evaluate $f(x)$ is required. Roughly speaking, it uses the secant rule unless bisection appears advantageous.

Normal input requires a continuous function $f(x)$ and arguments B and C for which $f(B)f(C) < 0$. Throughout the computation B and C will be the names of the endpoints of an interval which is decreasing in length and which preserves the property $f(B)f(C) < 0$. The values of B and C are always interchanged if necessary so that $|f(B)| \leq |f(C)|$ holds; for this reason we regard B as the better approximation to a root.

The convergence test is based on a mixed relative-absolute error test. The test asks whether

$$\left|\frac{C - B}{2}\right| \leq |B| \times \text{RELERR} + \text{ABSERR}. \tag{8}$$

If the input parameter RELERR is set to zero, this tests whether the interval containing a root has length no more than 2 × ABSERR. Since then the midpoint $M = \dfrac{C + B}{2}$ is no further than ABSERR from a root, this is a pure absolute error test. If the parameter ABSERR is set to zero and if the test were

$$\left|\frac{C - B}{2}\right| \leq |M| \times \text{RELERR},$$

the test would be a pure relative error test for the approximate root M. The test actually uses B instead of M because it should be the better approximate root. A fortuitous computed function value of *exactly* zero causes the code to return control to the calling program. Our version of the algorithm counts the function evaluations and quits after 500 are made. These last two situations may cause a return without the convergence test being satisfied.

Unless there is a reason to do otherwise, the secant rule is used. A variable A is initialized to C. The two variables A, B are the two iterates to be used by the secant rule. Using this rule, one calculates

$$D = B - f(B)\frac{B - A}{f(B) - f(A)}.$$

This computation is done cautiously to prevent unnecessary overflow and divide checks. At the same time, sudden large changes are rejected. Before actually dividing, one implicitly tests that the resulting D will lie in the interval $[B, C]$ which is known to contain a root. In point of fact, it is subjected to the more stringent test that D lie in $[B, M]$, on the grounds that B ought to be a better approximation to the root than C and, if the secant rule is working properly, D ought to be closer to B than to C. If D does not lie in $[B, M]$, it is rejected and one uses the midpoint M as the iterate.

There are circumstances when the above test is passed but it is still undesirable to use the secant rule. We prefer to reject values of D which are "too" close to B and always to move a minimum distance away from the last iterate. The quantity $|B| \times \text{RELERR} + \text{ABSERR}$ is called TOL in the code. If $|D - B| < \text{TOL}$, then the value $B + \text{TOL} \times \text{sign}(C - B)$ is used instead of D. This choice of minimum change cannot cause one to leave the interval $[B, C]$, since $|B - C| > 2 \times \text{TOL}$ (or else we would have already converged).

Even this use of a minimum change can allow the code to compute a great many iterates. There are circumstances in which C is fixed and B creeps up on a root by a change TOL each time. If one is requesting high accuracy, convergence can be long delayed. We have modified Dekker's algorithm so that the length of the interval $[B, C]$ is monitored. If four iterations have not resulted in a reduction by a factor of 1/8, the code bisects until it is. Thus, our version of the algorithm reduces an interval containing a root by a factor of 1/8 in a maximum of seven function evaluations.

In summary, if D lies outside $[B, M]$ or if the overall reduction in interval length has been unsatisfactory, we bisect. If D is "too" close to B, we use a minimum change of TOL. Otherwise the secant rule is used.

After the decision as to what will be used for the next iterate has been made, the iterate is explicitly computed and replaces B; the old B replaces A. Depending on the sign of the function value for the new B, the old C is kept or is replaced by the old B. The choice is made so that $f(B)f(C) < 0$. This is always possible unless $f(B) = 0$, in which case the code exits.

If one enters the code with normal input [$f(x)$ continuous, $f(B)f(C)$

$< 0]$, then normal output permits one to say the following. The output values of B and C satisfy the test (8). The *computed* function values $f(B)$ and $f(C)$ have opposite signs and $|f(B)| \le |f(C)|$. There is either a root of $f(x) = 0$ in the interval $[B, C]$ or else one of the endpoints is as close to a root as the precision permits. Why is this last statement true? If the signs of the computed $f(B)$ and $f(C)$ are correct, then because $f(x)$ is continuous there is a root in the interval. If one of the signs is incorrect, then we are at limiting precision and the argument is as close to a root as we may hope to obtain using this precision.

It is important for the user to understand what happens in the code in certain abnormal situations. If on input $f(B)f(C) > 0$, we do not know whether there are roots in the interval at all. The code bisects, always choosing the interval which has the smaller function values at its ends. If at any time a sign change is detected, it then proceeds in the usual way. If the length of the interval meets the convergence test but there is no sign change, this is reported to the user. The code may have computed a root of even multiplicity. If not, there *appears* to be no root in the interval.

The other abnormal situation of importance is when $f(x)$ is not continuous because it has poles. We say that α is a pole of order $m > 0$ if $1/f(x)$ has α as a zero of multiplicity m. For x near α we can write

$$f(x) = (x - \alpha)^{-m} g(x)$$

with $g(x)$ continuous near α and $g(\alpha) \neq 0$. Our code can converge to a pole of odd order just like a zero of odd order. Overflow may well occur; if it does not, one is still warned about a pole (usually) since the code compares $f(B)$ to the function values of the original input. If $f(B)$ is larger, then one almost certainly has a pole and a warning is given to the user. We shall shortly see that problems with poles are not infrequent.

Example. In Chapter II.2 we discussed a problem which required the solution of

$$\varphi(\lambda) = \frac{x^2}{a^2 + \lambda} + \frac{y^2}{b^2 + \lambda} + \frac{z^2}{c^2 + \lambda} - 1 = 0$$

for its smallest positive root. The particular values of the parameters used there were $x = y = z = 50$ and $a = 1$, $b = 2$, $c = 100$. A rough sketch of φ for $a^2 < b^2 < c^2$ is pictured in Figure II.7. This arises from looking at the behavior of φ as $\lambda \to -a^2, -b^2, -c^2, +\infty$, and $-\infty$. The portion of interest to us is between $-a^2$ and $+\infty$. Since as $\lambda \to -a^2$ from the right, $\varphi(\lambda) \to +\infty$ and since as $\lambda \to +\infty$, $\varphi(\lambda) \to -1$, the continuous function φ must have a zero larger than $-a^2$. Differentiating $\varphi(\lambda)$ gives

$$\varphi'(\lambda) = -\frac{x^2}{(a^2 + \lambda)^2} - \frac{y^2}{(b^2 + \lambda)^2} - \frac{z^2}{(c^2 + \lambda)^2} < 0.$$

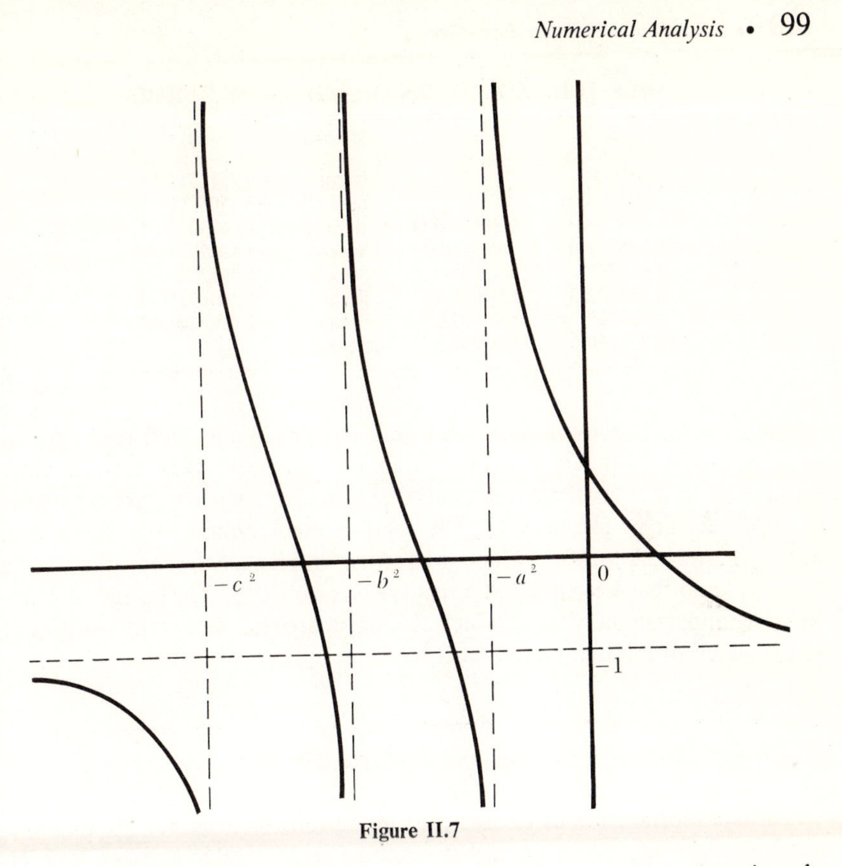

Figure II.7

This says several things. First, since $\varphi(\lambda)$ is strictly decreasing, there is only one zero λ_0 greater than $-a^2$. Second, the root is simple since $\varphi'(\lambda_0) \neq 0$.

An interesting fact about this problem is that

$$P(\lambda) = -(a^2 + \lambda)(b^2 + \lambda)(c^2 + \lambda)\varphi(\lambda)$$

is a cubic polynomial. Since any root of $\varphi(\lambda) = 0$, such as λ_0, also satisfies $P(\lambda_0) = 0$, we can apply some bounds on the relative error of an approximate zero of a polynomial developed in problems 7 and 8 of the exercises.

The equation $\varphi(\lambda) = 0$ was solved using the code ZEROIN with relative and absolute error requests of 0 and 10^{-6} respectively. Poor initial values of B and C were used to show the excellent convergence. Table II.1 shows successive values of B and C and tells which method was used by ZEROIN in computing B. Here, as in other tables in this text, we write numbers like 9.998797×10^3 as 9.998797(3).

It is rather interesting to note that the code terminated because the last value of B gave a computed value of exactly zero for $\varphi(B)$. This is especially true because B is also supposed to be a root of the equation $P(\lambda) = 0$. To use the error bounds of problems 7 and 8 we had to compute $P(B)$ and $P'(B)$ in double precision. The residual $P(B) = -5.389920 \times 10^5$! This dramatically illustrates the difference between numerical computation and mathematical computation. The value B is supposed to be a root of $\varphi(\lambda) = 0$

TABLE II.1. SOLUTION OF $\varphi(\lambda) = 0$ BY ZEROIN

B	C	*Method*	$\phi(B)$
1(4)	0	Input	−3.751251(−1)
9.998797(3)	0	Secant	−3.750576(−1)
4.999398(3)	9.998797(3)	Bisect	1.662931(−1)
6.535121(3)	4.999398(3)	Secant	−8.400297(−2)
6.019707(3)	4.999398(3)	Secant	−1.368225(−2)
5.919422(3)	6.019707(3)	Secant	1.360893(−3)
5.928492(3)	5.919422(3)	Secant	−1.978874(−5)
5.928359(3)	5.919422(3)	Secant	0

and $P(\lambda) = 0$. Exact computation would have $\varphi(B)$ and $P(B)$ both zero or both non-zero. Numerical computation has one value zero but the other quite large. This very large residual is due to scaling, which we ask the reader to think about in problem 1. The relative error bound of problem 8 is approximately 3.2×10^{-6}, which is excellent in view of the facts that it is a "bound" and that we requested a relative error of 10^{-6}. The bound of problem 7 is approximately 0.16, which is almost useless. Generally we do not know which of the two bounds will be better.

Condition, Limiting Precision, and Multiple Roots

It is important for us to ask what limitations on accuracy are imposed by our arithmetic. Since we seek a machine representable number $\bar{\alpha}$ which makes $f(\bar{\alpha})$ as nearly zero as possible, we are working in a situation when the details of the computation of f, the machine word length, and the roundoff characteristics may play a dominant role. We have remarked that the computed function values may vary erratically in an interval about the zero. As a trivial example, consider the polynomial $(x - 1)^3$ which we evaluate in the form $((x - 3)x + 3)x - 1$ in three decimal chopped floating point arithmetic. For $x = 1.00, 1.01, \ldots, 1.17$ the computed function values are exactly zero (!) with the exception of the value $+0.0100$ at $x = 1.01, 1.11, 1.15$. For $x = 1.18, 1.19, \ldots, 1.24$ all function values are $+0.0200$ except for a value of exactly zero at $x = 1.20$ and a value of $+0.0100$ at $x = 1.23$. The reader might enjoy evaluating the function for x values less than one to explore this phenomenon. It is clear that these erratic values can cause unpredictable and unstable behavior on the part of the secant rule. Evaluating the derivative shows that Newton's method can be unstable here, too.

What effect on the accuracy of a root does inaccuracy in the function values have? To get some feeling for this, suppose that the routine which computes $f(x)$ actually returns a value $\bar{f}(x)$, and for x a machine number near a root α, the best we can say is that

$$|f(x) - \bar{f}(x)| \leq \epsilon$$

for a suitable ϵ. Suppose that z is a machine number and $\bar{f}(z) = 0$. How much in error can z be? If α is of multiplicity m, then

$$f(z) \doteq (z - \alpha)^m \frac{f^{(m)}(\alpha)}{m!}.$$

Since it is possible for $f(z)$ to be as large as ϵ, we could have

$$\pm\epsilon \doteq (z - \alpha)^m \frac{f^{(m)}(\alpha)}{m!},$$

so it is possible that

$$|z - \alpha| \doteq \epsilon^{1/m} \left| \frac{m!}{f^{(m)}(\alpha)} \right|^{1/m}.$$

For small ϵ and $m > 1$ the term $\epsilon^{1/m}$ is much larger than ϵ, so there is a serious loss of accuracy. The other factor plays a role, but generally we must consider multiple roots to be ill conditioned (sensitive to errors in f). The ill conditioning of the root of multiplicity 3 in $(x - 1)^3 = 0$ is evident in the example above. We saw that $x = 1.20$ led to a function value of exactly zero, and this is certainly a poor approximation to the root at 1.00.

Even when $m = 1$, the root may be poorly determined. As one might expect, clusters of roots "look" like multiple roots but, even when well separated, roots can still be poorly determined. A famous example given by Wilkinson [45] shows this. We shall forego the analysis and report what happens. Consider the polynomial equation

$$(x - 1)(x - 2) \cdots (x - 19)(x - 20) = 0$$

which has the roots $1, 2, \ldots, 19, 20$. These roots are obviously simple and well separated. The coefficient of x^{19} is -210. If this coefficient is changed by a small amount such as 2^{-23} to become -210.000000119, the roots become those shown in Table II.2.

Notice that five pairs of roots have become complex with imaginary parts of substantial magnitude. There is really no remedy for this ill conditioning except to use more figures in the computations. Often it is fairly easy to use double precision in the function subroutines, and this should probably be done.

Multiple roots are awkward not only because of their ill conditioning but for other reasons, too. Except for bisection, the methods we have presented slow down drastically for multiple roots. Bisection cannot compute roots of even multiplicity because there is no sign change. If the derivative $f'(x)$ is available, one can do something about this. If

$$f(x) = (x - \alpha)^m g(x) \quad \text{near} \quad \alpha,$$

TABLE II.2. Roots of $(x-1)(x-2)\cdots(x-19)(x-20) - 2^{-23}x^{19} = 0$

1.00000 0000	6.00000 6944
2.00000 0000	6.99969 7234
3.00000 0000	8.00726 7603
4.00000 0000	8.91725 0249
4.99999 9928	
10.09526 6145	±0.64350 0904 *i*
11.79363 3881	±1.65232 9728 *i*
13.99235 8137	±2.51883 0070 *i*
16.73073 7466	±2.81262 4894 *i*
19.50243 9400	±1.94033 0347 *i*
20.84690 8101	

then

$$f'(x) = (x - \alpha)^{m-1}G(x) \quad \text{near} \quad \alpha$$

where

$$G(x) = mg(x) + (x - \alpha)g'(x)$$

and

$$G(\alpha) = mg(\alpha) \neq 0.$$

This says that zeros of $f(x)$ of even multiplicity are zeros of $f'(x)$ of odd multiplicity, so one could use ZEROIN to compute them. However, we could notice that

$$u(x) = \frac{f(x)}{f'(x)} = (x - \alpha)H(x) \quad \text{near} \quad \alpha$$

where

$$H(x) = g(x)/G(x), \qquad H(\alpha) = 1/m \neq 0,$$

so that $u(x)$ has only a simple zero. It is desirable to work with $u(x)$ because it speeds up ZEROIN and allows the code to compute all zeros. The only real difficulty is that a root of $f'(x) = 0$ might not be a root of $f(x) = 0$. Such a root leads to a pole of $u(x)$. We have mentioned that this does not pose a serious problem for ZEROIN, but we do have to be careful about overflow and unintentionally converging to an odd order pole.

EXERCISES

1. The residual of an alleged root r of $f(x) = 0$ is $f(r)$. One often sees the statement that a residual is "small," so the approximate root must be "good." Is this reliable? What role does scaling play?

Implement the ZEROIN code and use it where applicable to solve the following problems. In all cases try to decide whether the roots are simple or multiple, since this affects how easily and how well they are computed.

2. In the *Theory of Elastic Stability* (McGraw-Hill, New York, 1961 p. 151), Timoshenko and Gere derive an equation determining the critical load for columns with batten plates. Suitable values of the physical parameters for experiments performed by Timoshenko lead to the problem

$$\frac{1}{180} = \frac{1 - \cos(\pi/10)}{\cos(\pi/10) - \cos z} \cdot \frac{\sin z}{z}$$

and the smallest positive root is desired. Make a rough sketch of the function to get an idea where the root is. (Consider $z = \pi/10$ and π).

3. In J. Chem. Ed. *37* (1960), p. 422, Eberhardt and Sweet consider an equation for the temperature T at which o-toluidine has a vapor pressure of 500 mm Hg. In degrees absolute, T satisfies

$$21.1306 - \frac{3480.3}{T} - 5.081 \log_{10} T = 0.$$

A desk calculator and a table of logarithms quickly shows that T is between 300 and 600 degrees.

4. In the text by Wing, *An Introduction to Transport Theory* (Wiley, New York, 1962), the study of neutron transport in a rod leads to a transcendental equation which has roots related to the critical lengths. For a rod of length ℓ the equation is

$$\cot(\ell x) = \frac{x^2 - 1}{2x}.$$

Make a rough sketch of the two functions to get an idea of where they intersect to yield roots. For $\ell = 1$, determine the smallest positive root.

5. Volterra has written differential equations describing the problem of growth in two populations conflicting with one another. H. T. Davis, in *Introduction to Nonlinear Differential and Integral Equations* (Dover, New York, 1962), discusses the semi-analytical solution of these differential equations. It is necessary to solve nonlinear equations of the form

$$xe^{-x} = \gamma$$

where γ is a given positive number. Discuss the number and multiplicity of solutions of this equation if $\gamma < 1/e$. Davis gives a series for the smallest positive root,

$$x = \gamma + \gamma^2 + \tfrac{3}{2}\gamma^3 + \cdots + \frac{n^{n-1}}{n!}\gamma + \cdots.$$

For $\gamma = 0.06064$, solve this problem by Newton's method and compare with the series result. If you guess $x = 0$, what is the first approximation to the root?

6. The text by P. David and J. Voge, *Propagation of Waves* (Pergamon Press, New York, 1969), derives on p. 65 a cubic equation for a parameter s in the

context of corrections for the earth's curvature in the interference zone. The equation

$$s^3 - \tfrac{3}{2}s^2 - \frac{s}{2}\left(\frac{1+u}{v^2} - 1\right) + \frac{1}{2v^2} = 0$$

depends on two parameters, u and v, which are obtained from the heights of the towers, the distance between stations, and the radius of the earth. Representative values are $v = 1/291$, $u = 30$. The smallest positive root is the one of interest, but calculate them all. The residuals of the larger roots are quite large. Are they inaccurate? Compare with problem 1. Use the computable error bounds of problems 7 and 8 to bound the errors of the roots.

The next two problems derive error bounds for an approximate root σ (real or complex) of the polynomial equation

$$P(x) = x^n + a_{n-1}x^{n-1} + \cdots + a_1 x + a_0 = 0.$$

In each case we require an accurate value of $P(\sigma)$. Since root solvers may make this residual about as small as it can be in single precision, it is necessary to compute $P(\sigma)$ in double precision. Let $r_1, r_2, \ldots, r_n$ be the roots of $P(x) = 0$.

7. The theory of equations [44] tells us that $P(x)$ can be factored in the form

$$P(x) = (x - r_1)(x - r_2) \cdots (x - r_n).$$

Show that

$$a_0 = (-1)^n r_1 r_2 \cdots r_n$$

and then that

$$\left|\frac{P(\sigma)}{a_0}\right| \geq \min_j \left|\frac{\sigma - r_j}{r_j}\right|^n.$$

This implies that

$$\min_j \left|\frac{\sigma - r_j}{r_j}\right| \leq \left|\frac{P(\sigma)}{a_0}\right|^{1/n}$$

which says that there is some zero which is approximated with a relative error of no more than

$$\left|\frac{P(\sigma)}{a_0}\right|^{1/n}$$

This bound gives a poor result if it is applied to a large zero when the polynomial has small zeros. Explain why.

8. Show that

$$\frac{P'(\sigma)}{P(\sigma)} = \sum_{j=1}^{n} \frac{1}{\sigma - r_j}$$

by differentiating $\ln P(x)$. This then implies that

$$\left|\frac{P'(\sigma)}{P(\sigma)}\right| \leq n \cdot \frac{1}{\min_j |\sigma - r_j|}$$

and

$$\min_j |\sigma - r_j| \leq n \left| \frac{P(\sigma)}{P'(\sigma)} \right| .$$

This is an absolute error bound, but we get the following relative error bound easily:

$$\min_j \left| \frac{\sigma - r_j}{\sigma} \right| \leq n \left| \frac{P(\sigma)}{\sigma P'(\sigma)} \right| .$$

What is the relation between this bound and Newton's method?

9. Write a code like ZEROIN based upon bisection and Newton's method. Are there advantages to using Newton's method?

10. Modify ZEROIN so as to supply $f'(x)$ in addition to $f(x)$ and to compute roots via the function $u(x) = f(x)/f'(x)$ as described in the text. This makes the modified code faster for multiple roots and permits one to compute roots of even multiplicity. Consider the possibility of detecting poles without being thrown off the system by overflow and of taking appropriate action so as always to compute a root.

11. How are simple and multiple roots distinguished graphically? Interpret graphically how well the roots are determined. Compare with problem 1.

12. In Chapter II.2, problem 10, a temperature distribution problem is solved in terms of a series. For the solution it is necessary to compute positive roots of

$$2xJ_1(x) - J_0(x) = 0$$

where $J_0(x)$ and $J_1(x)$ are Bessel functions of the first kind of order zero and one respectively. Compute the three smallest positive roots.

13. ZEROIN is very efficient. Suppose $|B - C| = 10^{10}$ on input. What is the *maximum* number of function evaluations required to locate a root to an absolute accuracy of 10^{-5}?

14. In trying to solve the equations of radiative transfer in semi-infinite atmospheres, one encounters the nonlinear equation

$$\omega_0 = \frac{2k}{\ln\left(\dfrac{1 + k}{1 - k}\right)}$$

where the number ω_0, $0 < \omega_0 < 1$, is called an albedo. Show that for fixed ω_0 if k is a root, so is $-k$, and that there is a unique value of k with $0 < k < 1$ satisfying the equation. For $\omega_0 = 0.25$, 0.50, and 0.75, find the corresponding positive k values. Make some sketches to help you locate the roots.

15. A wire weighs 0.518 lb. per ft. and is suspended between two towers of equal height (at the same level) and 500 ft. apart. If the sag in the wire is 50 ft., find

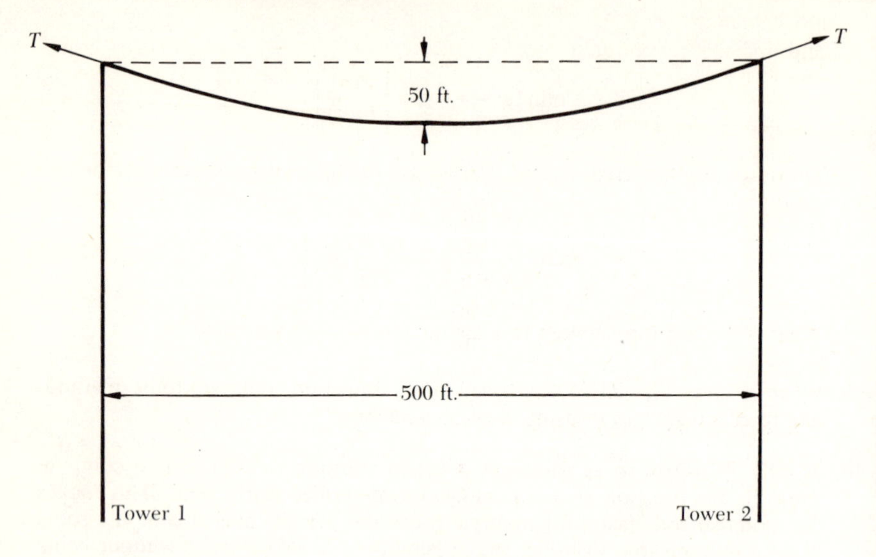

Figure II.8

the maximum tension T in the wire. (See Figure II.8.) The appropriate equations to be solved are

$$c + 50 = c \cosh \frac{500}{2c}$$

$$T = 0.518(c + 50).$$

16. A semi-infinite medium is at a uniform initial temperature $T_0 = 70°\text{F}$. For time $t > 0$, a constant heat flux density $q = 300$ BTU/hr. sq. ft. is maintained on the surface $x = 0$. Knowing the thermal conductivity $k = 1.0$ BTU/hr. ft.°F and the thermal diffusivity $\alpha = 0.04$ sq. ft./hr., the resulting temperature $T(x, t)$ is given by

$$T(x, t) = T_0 + \frac{q}{k}\left[2\sqrt{\frac{\alpha t}{\pi}}\, e^{-x^2/4\alpha t} - x\left(1 - \operatorname{erf}\left(\frac{x}{2\sqrt{\alpha t}}\right)\right)\right]$$

where

$$\operatorname{erf}(y) = \frac{2}{\sqrt{\pi}} \int_0^y e^{-z^2}\, dz$$

is the error function. Find the times t required for the temperature at distances $x = .1, .2, \ldots, .5$ to reach a preassigned value $T = 100°\text{F}$. The function ERF(Y) is supplied in the FORTRAN library on the IBM 360 system, and nearly all computing centers provide a code if it is not available in the FORTRAN library.

17. Establish the relationship (7).

18. For turbulent flow of fluid in a smooth pipe, the equation

$$1 = \sqrt{c_f}(-0.4 + 1.74 \ln(\text{Re}\sqrt{c_f}))$$

governs the relationship between the friction factor c_f and the Reynold's number Re. Compute c_f for Re $= 10^4, 10^5, 10^6$.

19. In the convergence proof for the secant rule, it was stated that if $\epsilon = \max(\epsilon_0, \epsilon_1) < 1$, then the inequality

$$\epsilon_{i+1} \leq \epsilon_i \cdot \epsilon_{i-1}$$

implied

$$\epsilon_i \leq \epsilon^{\delta_i}, \qquad \delta_i = \frac{1}{\sqrt{5}}\left[\left(\frac{1+\sqrt{5}}{2}\right)^{i+1} - \left(\frac{1-\sqrt{5}}{2}\right)^{i+1}\right].$$

Establish this.

20. The rectangular potential well pictured in Figure II.9 arises in the study of problems in quantum mechanics [29]. An analysis of the Schrödinger equation for a particle of mass m associated with this potential leads to discrete sets of

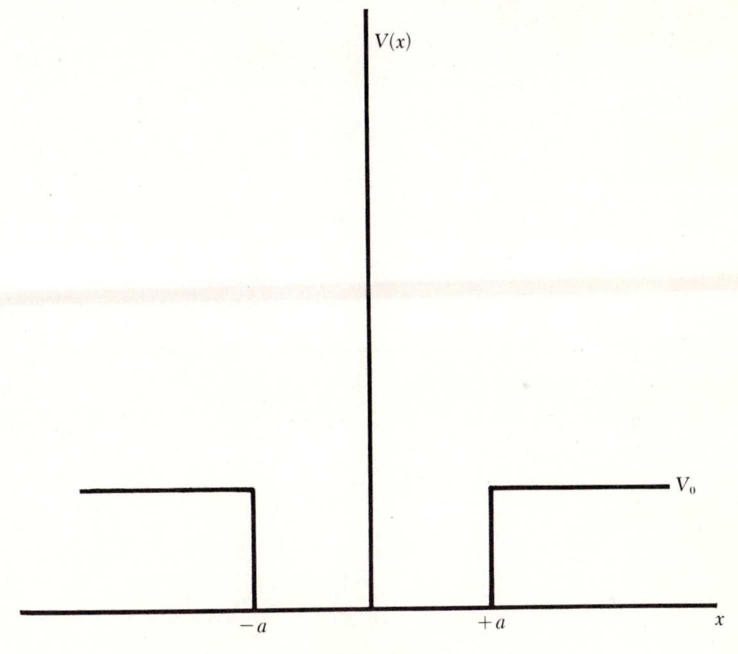

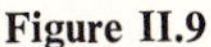

Figure II.9

values of the total energy E which are solutions of a pair of transcendental equations. One of these equations is

$$\cot\left(\frac{a}{\hbar}\sqrt{2mV_0}\sqrt{E/V_0}\right) = \sqrt{\frac{E/V_0}{1 - E/V_0}}$$

where

$$\hbar = \frac{h}{2\pi}, \qquad h = 6.625 \times 10^{-27} \text{ erg-sec}$$

is Planck's constant. Find the smallest positive value of E which satisfies this equation. Use the following data, which correspond to a simplified model of the hydrogen atom:

$$m = 9.109 \times 10^{-28} \text{ gm.},$$

$$V_0 = 2.179 \times 10^{-11} \text{ erg},$$

$$a = 5.292 \times 10^{-9} \text{ cm}.$$

4. ORDINARY DIFFERENTIAL EQUATIONS

Elements of the Theory

Ordinary differential equations are used to describe the most diverse problems in a wide range of disciplines. As a consequence, they appear in many different forms. We begin with what is apparently a rather special case, but which is really of great practical importance, namely the numerical solution of the first order ordinary differential equation

$$y' = f(x, y) \tag{1}$$

on a finite interval $a \leq x \leq b$. The function of two variables $f(x, y)$ is in general nonlinear, and we shall assume that it is continuous for $a \leq x \leq b$ and all y. By a solution $y(x)$ we mean a function of x with one continuous derivative for $a \leq x \leq b$ (we denote this by $y(x) \in C^1[a, b]$) which satisfies the equation

$$y'(x) = f(x, y(x))$$

for each x in the interval $[a, b]$. There can be many solutions, as the example

$$y' = 0$$

shows, since the constant function $y(x) \equiv C$ is a solution on any interval for any C. Most practical problems have only one solution, which is specified by an additional requirement. The typical problem for ordinary differential equations supplies the value of the desired solution at the initial point of the interval:

$$y(a) = A. \tag{2}$$

Equation (2) is called an initial condition, and the combination of (1) and (2) is called an initial value problem for an ordinary differential equation.

In elementary treatments of differential equations, the initial value problem has a unique solution which exists throughout the interval of interest and which can be obtained by analytical means (more familiarly called a "trick"). This good fortune does not persist as one begins to encounter more realistic problems; it is probably fair to say that for most problems which are not contrived, an analytical solution is impossible to obtain or is less satisfactory than a numerical solution. Matters are also complicated by the fact that solutions can exhibit rather unpleasant behavior.

For example, let us look at the problem

$$y' = \sqrt{|y|},$$
$$y(0) = 0$$

on some interval $[0, b]$. There is the obvious solution $y(x) \equiv 0$, but there are other solutions, even a whole family of them! Let c be any constant such that $0 \leq c \leq b$; then

$$\begin{aligned} y(x) &\equiv 0, \quad \text{if} \quad 0 \leq x \leq c, \\ &= \tfrac{1}{4}(x - c)^2, \quad \text{if} \quad c \leq x \leq b \end{aligned}$$

is a solution of the initial value problem. The reader should verify that this function $y(x) \in C^1[0, b]$ and satisfies the differential equation and the initial condition. Another difficulty is exemplified by

$$y' = |y|^{3/2},$$
$$y(0) = 1$$

when we want a solution on the interval $[0, 3]$. The differential equation has a solution satisfying the initial condition, namely

$$y(x) = \left(1 - \frac{x}{2}\right)^{-2},$$

but it "blows up" at $x = 2$ and so does not exist throughout $[0, 3]$. Problems which have more than one solution are quite difficult to solve numerically, and we shall not discuss them here. Problems with solutions which "blow up" place a special burden on a numerical procedure, though we might well expect a general purpose code to compute such solutions until overflow occurs in the computations. These non-routine difficulties will be excluded at the level of the theory to be developed.

There is a simple condition which excludes unpleasant behavior of solutions and suffices to establish effective numerical procedures. It is formulated in terms of how fast f can grow in the y variable; specifically, we shall assume that $f(x, y)$ satisfies a Lipschitz condition in y. By this we mean that for *all* x in the interval $[a, b]$ and for *all* u, v,

$$|f(x, u) - f(x, v)| \leq L\,|u - v| \tag{3}$$

with L a constant, hereafter called a Lipschitz constant. The inequality assumes a more familiar form if f has a continuous partial derivative in its second variable, for then

$$|f(x, u) - f(x, v)| = \left|\frac{\partial f}{\partial y}(x, w)\right|\,|u - v|$$

for some w between u and v. If $\partial f/\partial y$ is bounded in magnitude for all arguments, then f satisfies a Lipschitz condition, and a suitable Lipschitz constant is any finite constant L such that

$$\left|\frac{\partial f}{\partial y}(x, w)\right| \leq L$$

for all x in $[a, b]$ and all w. If the partial derivative is not bounded, it is not hard to show that the inequality (3) cannot hold for all u, v and all x in $[a, b]$, so f does not satisfy a Lipschitz condition at all.

Examples. The function $f(x, y) = x^2 \cos^2 y + y \sin^2 x$ defined for $|x| \leq 1$ and all y satisfies a Lipschitz condition. The constant $L = 3$ since differentiating with respect to y gives

$$\frac{\partial f}{\partial y} = -2x^2 \cos y \sin y + \sin^2 x$$

and

$$|\partial f/\partial y| \leq 2 \times 1 \times 1 \times 1 + 1 = 3$$

for the range of x allowed. The functions $f(x, y) = \sqrt{|y|}$ and $f(x, y) = |y|^{3/2}$ do not satisfy Lipschitz conditions, since they have continuous partial derivatives for $y > 0$ which are not bounded as $y \to 0$ and $y \to +\infty$ respectively:

$$\frac{\partial f}{\partial y} = \frac{1}{2\sqrt{y}}, \quad = \tfrac{3}{2}\sqrt{y}.$$

An important special case is that of a linear differential equation, meaning that $f(x, y)$ has the form $f(x, y) = g(x)y + h(x)$. The function $f(x, y)$ being continuous is then equivalent to $g(x)$ and $h(x)$ being continuous in x. Because

$$\frac{\partial f}{\partial y} = g(x),$$

and because a continuous function $g(x)$ is bounded in magnitude on any finite interval $[a, b]$, a linear equation is Lipschitzian in nearly all cases of practical interest. As an example, let us consider Dawson's integral which is the function

$$y(x) = e^{-x^2} \int_0^x e^{t^2}\, dt.$$

The reader should verify that it is a solution of the initial value problem for the linear differential equation

$$y' = 1 - 2xy,$$

$$y(0) = 0.$$

On the interval $[0, b]$ for any $b \neq 0$, the function $f(x, y) = 1 - 2xy$ is continuous and Lipschitzian, with Lipschitz constant $L = 2\,|b|$.

The reason we are interested in Lipschitzian functions is the following theorem, which is proved in [11].

Theorem 1. Let $f(x, y)$ be continuous for $a \leq x \leq b$ and all y and satisfy a Lipschitz condition. Then for any number A, the initial value problem $y' = f(x, y)$, $y(a) = A$ has a unique solution $y(x)$ which is defined for all x in the (finite) interval $[a, b]$.

For most differential equations ordinarily encountered, the equation can be easily modified to make the function f Lipschitzian. This modification is necessary for the theory we develop but does not have to be made in practice. More will be said about this in the "miscellaneous" section, (page 135).

So far we have spoken of a single equation in a single unknown $y(x)$. More commonly, one encounters many unknowns. By a system of n first order equations in n unknowns, we mean

$$\begin{aligned} Y_1' &= F_1(x, Y_1, Y_2, \ldots, Y_n), \\ Y_2' &= F_2(x, Y_1, Y_2, \ldots, Y_n), \\ &\;\cdot \\ &\;\cdot \\ &\;\cdot \\ Y_n' &= F_n(x, Y_1, Y_2, \ldots, Y_n). \end{aligned} \tag{4}$$

Along with (4) we associate the initial conditions

$$\begin{aligned} Y_1(a) &= A_1, \\ Y_2(a) &= A_2, \\ &\;\cdot \\ &\;\cdot \\ &\;\cdot \\ Y_n(a) &= A_n. \end{aligned} \tag{5}$$

This can be written in tidy fashion using vector notation. Let

$$\boldsymbol{Y}(x) = \begin{bmatrix} Y_1(x) \\ Y_2(x) \\ \cdot \\ \cdot \\ \cdot \\ Y_n(x) \end{bmatrix}, \qquad \boldsymbol{A} = \begin{bmatrix} A_1 \\ A_2 \\ \cdot \\ \cdot \\ \cdot \\ A_n \end{bmatrix}, \qquad \boldsymbol{F}(x, \boldsymbol{Y}) = \begin{bmatrix} F_1(x, \boldsymbol{Y}) \\ F_2(x, \boldsymbol{Y}) \\ \cdot \\ \cdot \\ \cdot \\ F_n(x, \boldsymbol{Y}) \end{bmatrix},$$

then (4) and (5) become

$$Y' = F(x, Y),$$

$$Y(a) = A.$$

We again refer to the combination of (4) and (5) as an initial value problem. Using vector notation makes the case of n unknowns look like the case of one unknown. One of the fortunate aspects of the theory of the initial value problem is that the theory for a system of n first order equations is essentially the same as for a single one. Proofs for systems just introduce vectors, and things like the length of a vector instead of the absolute value of a scalar, into the proofs for a single equation. For the vector function $F(x, Y)$ to satisfy a Lipschitz condition, it suffices that each $F_i(x, Y_1, Y_2, \ldots, Y_n)$ satisfy a Lipschitz condition with respect to each variable Y_j; that is, there are constants L_{ij} such that

$$|F_i(x, Y_1, \ldots, Y_{j-1}, u, Y_{j+1}, \ldots, Y_n) - F_i(x, Y_1, \ldots, Y_{j-1}, v, Y_{j+1}, \ldots, Y_n)| \le L_{ij}\, |u - v| \quad \text{for each } i, j.$$

With this, the natural analog of Theorem 1 holds. Since the theory of numerical methods for a system of equations is also essentially the same as for a single equation, we shall content ourselves with treating the case of a single unknown in detail and just state the analog for systems.

Most computer codes accept differential equations in the standard form (4) and (5), but the equations often arise in different forms. For example, one is often interested in second order equations, that is, equations of the form

$$y'' = g(x, y, y').$$

The definition of a solution is the obvious extension of the first order case. Suitable initial conditions are $y(a) = A_1$, $y'(a) = A_2$. In this problem there is one unknown quantity, $y(x)$. From it we can, in principle, obtain $y'(x)$ as needed. A new problem will be formed which involves two unknown quantities, $Y_1(x)$ and $Y_2(x)$, but which is in the standard form (4). The idea is to approximate independently $y(x)$ by $Y_1(x)$ and $y'(x)$ by $Y_2(x)$. Suppose we could find suitable $Y_1(x) = y(x)$ and $Y_2(x) = y'(x)$. Then we would have

$$Y_1'(x) = y'(x) = Y_2(x),$$

$$Y_2'(x) = y''(x) = g(x, y(x), y'(x)) = g(x, Y_1(x), Y_2(x)).$$

This suggests considering the problem

$$Y_1' = Y_2,$$

$$Y_2' = g(x, Y_1, Y_2).$$

This is in standard form, and the theory may be applied to conclude the existence of unique functions $Y_1(x)$ and $Y_2(x)$ which satisfy initial conditions

$$Y_1(a) = A_1,$$
$$Y_2(a) = A_2.$$

Then $y(x) = Y_1(x)$ is the solution of the original problem. One of the equations says that $y'(x) = Y_1'(x) = Y_2(x)$; hence, the other equation is

$$y''(x) = Y_2'(x) = g(x, Y_1(x), Y_2(x)) = g(x, y(x), y'(x))$$

and similarly the initial conditions are satisfied.

More generally, an nth order equation in one unknown

$$y^{(n)} = g(x, y, y', \ldots, y^{(n-1)}),$$
$$y(a) = A_1, y'(a) = A_2, \ldots, y^{(n-1)}(a) = A_n$$

can be put into standard form via the n unknowns $Y_1(x) = y(x)$, $Y_2(x) = y'(x), \ldots, Y_n(x) = y^{(n-1)}(x)$ and

$$\begin{aligned} F_1(x, Y_1, Y_2, \ldots, Y_n) &= Y_2, \\ F_2(x, Y_1, Y_2, \ldots, Y_n) &= Y_3, \\ &\vdots \\ F_{n-1}(x, Y_1, Y_2, \ldots, Y_n) &= Y_n, \\ F_n(x, Y_1, Y_2, \ldots, Y_n) &= g(x, Y_1, Y_2, \ldots, Y_n). \end{aligned}$$

This is the usual way to go to a system of first order equations, but there are many other ways to do it. For example, if we are interested in the solution of

$$y''(x) + xy'(x) + x^2y(x) = 0,$$

we could go to the system

$$Y_1' = Y_2,$$
$$Y_2' = -x^2Y_1 - xY_2$$

or

$$Y_1' = e^{-x^2/2}Y_2,$$
$$Y_2' = -x^2e^{x^2/2}Y_1$$

or

$$Y_1' = -\frac{x}{2}Y_1 + Y_2,$$
$$Y_2' = (\tfrac{1}{2} - 3x^2/4)Y_1 - \frac{x}{2}Y_2.$$

In each case $Y_1(x) = y(x)$; this is easily verified for the first two systems, and the third is only slightly more difficult; we leave it as an exercise.

An important change of variables was introduced by Liénard in treating problems of nonlinear mechanics like

$$y''(x) + f(y(x))y'(x) + y(x) = 0$$

and similar equations known today by his name. We could, of course, make the usual choice $Y_1(x) = y(x)$, $Y_2(x) = y'(x)$ and consider the system

$$Y_1'(x) = Y_2(x),$$
$$Y_2'(x) = -f(Y_1(x))Y_2(x) - Y_1(x).$$

In many problems the function $f(t)$ is simple in form but only smooth in pieces. If we have an analytical expression for an indefinite integral $G(t)$ of the function $f(t)$, that is, $G'(t) = f(t)$, then we notice that

$$y''(x) + f(y(x))y'(x) + y(x) = \frac{d}{dx}(y'(x) + G(y(x))) + y(x).$$

So we use the Liénard variables $Y_1(x) = y(x)$, $Y_2(x) = y'(x) + G(y(x))$ and find

$$Y_1'(x) = Y_2(x) - G(Y_1(x)),$$
$$Y_2'(x) = -Y_1(x).$$

This system has smoother functions than our usual system, which we shall see makes it easier to solve numerically.

As a final example, consider the system of second order equations

$$u'' + 5v' + 7u = \sin x,$$
$$v'' + 6v' + 4u' + 3u + v = \cos x,$$
$$u(0) = 1, \qquad u'(0) = 2,$$
$$v(0) = 3, \qquad v'(0) = 4.$$

Let $Y_1(x) = u(x)$, $Y_2(x) = u'(x)$, $Y_3(x) = v(x)$, $Y_4(x) = v'(x)$. Then the equations are

$$Y_2' + 5Y_4 + 7Y_1 = \sin x,$$
$$Y_4' + 6Y_4 + 4Y_2 + 3Y_1 + Y_3 = \cos x$$

which we rearrange as

$$Y_1' = Y_2,$$
$$Y_2' = -7Y_1 - 5Y_4 + \sin x,$$
$$Y_3' = Y_4,$$
$$Y_4' = -3Y_1 - 4Y_2 - Y_3 - 6Y_4 + \cos x,$$

with initial conditions

$$Y_1(0) = 1, \qquad Y_2(0) = 2, \qquad Y_3(0) = 3, \qquad Y_4(0) = 4.$$

The main point here is that we have to introduce unknowns to handle derivatives up to one less than the highest appearing for each unknown in the original system. Since differential equations arise in the most diverse forms and combinations, the student is well advised to work all of the problems of this kind and to study their solution. One cannot do even meaningless computation until he can put his problem in a form acceptable to his code, so this is a very important matter.

A Simple Numerical Scheme

Let us again consider the initial value problem (1) and (2),

$$y' = f(x, y),$$
$$y(a) = A$$

on the interval $[a, b]$. The kind of numerical method we shall study generates a table of approximate values for $y(x)$. For the moment we suppose that the entries are for equally spaced arguments. Choose an integer N and let $h = (b - a)/N$. We seek approximations at the points $x_k = a + kh$ for $k = 0, 1, \ldots, N$. The expression $y(x_k)$ is always used for the solution of (1) and (2) evaluated at $x = x_k$, and y_k is always used for an approximation to $y(x_k)$.

A differential equation has no "memory." If we know the value $y(x_k)$, Theorem 1 applies to the problem

$$u' = f(x, u),$$
$$u(x_k) = y(x_k)$$

and says that the solution of this problem on the interval $[x_k, b]$ is just $y(x)$. (After all, $y(x)$ is *a* solution and the theorem says there is only one.) That is, the values of $y(x)$ for x prior to $x = x_k$ do not directly affect the solution of the differential equation for x after x_k. Some numerical methods have memory and some do not. The class of methods known as one-step methods have no memory; given y_k, there is a recipe for the value y_{k+1} which depends only on x_k, y_k, f, and h. Starting with the obvious initial value $y_0 = A$, a one-step method generates a table for $y(x)$ by repeatedly taking one step in x of length h to successively generate $y_1, y_2, \ldots$.

The simplest example of a one-step method is Euler's method. We shall study it because the details do not obscure the ideas, and so we can subsequently model our treatment of general one-step methods after it.

A Taylor's series expansion of $y(x)$ about $x = x_k$ gives

$$y(x_{k+1}) = y(x_k) + hy'(x_k) + \frac{h^2}{2} y''(\xi_k)$$

or

$$y(x_{k+1}) = y(x_k) + hf(x_k, y(x_k)) + \frac{h^2}{2} y''(\xi_k) \tag{6}$$

with $x_k < \xi_k < x_{k+1}$, provided $y(x) \in C^2[a, b]$. For small h, this is approximately

$$y(x_{k+1}) \doteq y(x_k) + hf(x_k, y(x_k)).$$

This suggests Euler's method:

$$\begin{aligned} y_0 &= A, \\ y_{k+1} &= y_k + hf(x_k, y_k) \qquad k = 0, 1, \ldots, N-1. \end{aligned} \tag{7}$$

As an example, let us tabulate Dawson's integral on [0, 0.5] using Euler's scheme with $h = 0.1$. Recall that Dawson's integral is the solution of the initial value problem

$$y' = 1 - 2xy,$$
$$y(0) = 0.$$

So, taking $y_0 = 0$, we see that

$$y_1 = 0 + 0.1 \times (1 - 2 \times 0 \times 0) = 0.1;$$

similarly,

$$y_2 = 0.1 + 0.1 \times (1 - 2 \times 0.1 \times 0.1) = 0.198.$$

Working on a desk calculator, we made the following table. The true values of the integral are taken from the *Handbook of Mathematical Functions* [1].

x_k	y_k	$y(x_k)$
0.	0.	0.
0.1	.10000	.09934
0.2	.19800	.19475
0.3	.29008	.28263
0.4	.37268	.35994
0.5	.44287	.42444

Euler's method can be used even if $f(x, y)$ is just continuous; however, for numerical purposes we shall usually need to assume that f has several partial derivatives. This guarantees that $y''(x)$ is bounded on $[a, b]$, and since we shall be interested in small h, equation (6) assures us that $y(x_k)$ "nearly" satisfies the recipe (7) for y_k. Subtracting (7) from (6) gives

$$y(x_{k+1}) - y_{k+1} = y(x_k) - y_k + h[f(x_k, y(x_k)) - f(x_k, y_k)] + \frac{h^2}{2} y''(\xi_k),$$

which is effectively a relationship between the error at x_k and the error at x_{k+1}. To be specific, let $e_k = y(x_k) - y_k$ and use the Lipschitz condition on f to arrive at

$$|e_{k+1}| \le |e_k| + hL\,|y(x_k) - y_k| + \frac{h^2}{2}\,|y''(\xi_k)|.$$

Letting

$$M_2 = \max_{a \le x \le b} |y''(x)|,$$

we obtain

$$|e_{k+1}| \le |e_k|\,(1 + hL) + \frac{h^2}{2} M_2 \qquad k = 0, 1, \ldots, N - 1. \tag{8}$$

Starting with an initial error of $|e_0| = |y(a) - y_0|$, this inequality bounds how much larger the error can be at x_{k+1} than it was at x_k. The idea of convergence is to bound the worst error that can arise as we step from $x_0 = a$ to $x_N = b$ and assert that it tends to zero as h does.

The first order of business is to see how rapidly inequality (8) permits the error to grow. We shall establish a more general result for later use. Suppose there are numbers $\delta > 0$ and $M > 0$ such that the sequence $d_0, d_1, \ldots$ satisfies

$$d_{k+1} \le (1 + \delta)\, d_k + M \qquad k = 0, 1, \ldots.$$

The case $k = 0$,

$$d_1 \le (1 + \delta)\, d_0 + M,$$

can be combined with the case $k = 1$ to obtain

$$d_2 \le (1 + \delta)\, d_1 + M \le (1 + \delta)^2 d_0 + M[1 + (1 + \delta)].$$

Similarly, one finds

$$d_3 \le (1 + \delta)\, d_2 + M \le (1 + \delta)^3 d_0 + M[1 + (1 + \delta) + (1 + \delta)^2].$$

At this point we might guess that

$$d_n \le (1 + \delta)^n d_0 + M[1 + (1 + \delta) + (1 + \delta)^2 + \cdots + (1 + \delta)^{n-1}]. \tag{9}$$

To prove this, we use induction. The inequality (9) certainly holds for $n = 1, 2, 3$. Suppose the inequality is true for the case $k = n$. Then

$$\begin{aligned} d_{n+1} &\leq (1 + \delta)\, d_n + M \\ &\leq (1 + \delta)^{n+1}\, d_0 + M[1 + (1 + \delta) + \cdots + (1 + \delta)^n] \end{aligned}$$

which establishes the result for $k = n + 1$, and completes the induction argument.

The inequality (9) may be put into a simpler form, which is stated as a lemma.

Lemma 1. Suppose there are numbers $\delta > 0$ and $M > 0$ such that the sequence $d_0, d_1, \ldots$ satisfies

$$d_{n+1} \leq (1 + \delta)\, d_k + M, \qquad k = 0, 1, \ldots.$$

Then for any $n \geq 0$

$$d_n \leq e^{n\delta} d_0 + M\, \frac{e^{n\delta} - 1}{\delta}. \tag{10}$$

PROOF. Using the identity

$$(x - 1) \sum_{p=0}^{n-1} x^p = \sum_{l=1}^{n} x^l - \sum_{p=0}^{n-1} x^p = x^n - x^0 = x^n - 1,$$

with $x = 1 + \delta$, we see that the right hand side of (9) can be rewritten in the form

$$(1 + \delta)^n d_0 + M\, \frac{(1 + \delta)^n - 1}{\delta}. \tag{11}$$

Taylor's expansion of the exponential function about zero gives for $\delta > 0$

$$e^\delta = 1 + \delta + \frac{\delta^2}{2} e^\zeta, \qquad 0 < \zeta < \delta.$$

It then follows that

$$1 + \delta \leq e^\delta$$

and

$$(1 + \delta)^n \leq e^{n\delta}.$$

So (11) is bounded by

$$e^{n\delta} d_0 + M\, \frac{e^{n\delta} - 1}{\delta}$$

which establishes (10).

Returning now to Euler's method, let us apply the lemma to (8) and so arrive at

$$|e_n| \leq e^{nhL}\,|e_0| + \frac{hM_2}{2L}(e^{nhL} - 1).$$

However, $nh = x_n - a$, so this is

$$|y(x_n) - y_n| \leq |y(x_0) - y_0|\,e^{L(x_n - a)} + \frac{hM_2}{2L}(e^{L(x_n-a)} - 1).$$

This inequality is the key to convergence statements. Of the several kinds of numerical convergence, we shall speak of uniform convergence, which is $\max\limits_{n=0,1,\ldots,N} |y(x_n) - y_n|$ tending to zero. Since $x_n - a \leq b - a$, the preceding inequality gives the bound

$$\max_{n=0,1,\ldots,N} |y(x_n) - y_n| \leq |y(x_0) - y_0|\,e^{L(b-a)} + \frac{hM_2}{2L}(e^{L(b-a)} - 1)$$

for the worst error in the table. Normally in convergence discussions we suppose that $|y(x_0) - y_0| = 0$, since we supply the exact initial value. Then we can say that the error is $0(h)$, which is a symbol for the error being bounded by a constant multiple of h. By taking h small enough we can reduce the error below any prescribed value.

Let us apply this bound to the table of Dawson's integral. In that example $a = 0$, $b = 0.5$, $L = 1$. To find a value for M_2 we note that since $y'(x) = 1 - 2xy(x)$,

$$\begin{aligned} y''(x) &= -2y(x) - 2xy'(x) \\ &= -2x + (4x^2 - 2)y(x), \end{aligned}$$

and

$$\max_{0 \leq x \leq 0.5} |y''(x)| \leq 1 + 2 \max_{0 \leq x \leq 0.5} |y(x)|.$$

But

$$y(x) = e^{-x^2}\int_0^x e^{t^2}\,dt = xe^{-x^2}e^{\xi^2}$$

where $0 < \xi < x$, from which we see that

$$0 \leq y(x) < x.$$

Thus, we can take $M_2 = 2$ and so

$$|y(x_k) - y_k| \leq 0.1(e^{x_k} - 1).$$

From our table we compare the true error and this bound to obtain

x_k	$\lvert y(x_k) - y_k \rvert$	bound
0	0	0
0.1	.00066	.01052
0.2	.00325	.02214
0.3	.00745	.03498
0.4	.01274	.04918
0.5	.01843	.06487

The bound is ordinarily of theoretical interest only, since usually we cannot obtain the quantities involved (mainly L and M_2) and since usually it is extremely pessimistic. The very nature of the bound supposes that the error grows as fast as it can but the error can decrease as well as increase.

The only way that a non-zero starting error $|y(x_0) - y_0|$ normally arises is from rounding A to a machine representable quantity. To better reflect the computational process, we might recognize this and even say we do not obtain $f(x_k, y_k)$ in a function subroutine, but rather obtain $f(x_k, y_k) + \epsilon_k$. Similarly, in computing $y_{k+1} = y_k + h[f(x_k, y_k) + \epsilon_k]$ we make an additional error ρ_k. The sequence generated computationally is then

$$y_{k+1} = y_k + hf(x_k, y_k) + h\epsilon_k + \rho_k.$$

Supposing that $|\rho_k| \leq \rho$ and $|\epsilon_k| \leq \epsilon$ for all $h \leq h_0$, a simple extension of the preceding analysis gives

$$\max |y(x_n) - y_n| \leq |y(x_0) - y_0| \, e^{L(b-a)} + \frac{e^{L(b-a)} - 1}{L}\left(\frac{hM_2}{2} + \epsilon + \frac{\rho}{h}\right).$$

One-Step Methods

Let us now consider general one-step methods and base our assumptions on the successful treatment of Euler's method. The recipe is to be of the form

$$\begin{aligned} y_0 &= A, \\ y_{k+1} &= y_k + h\Phi(x_k, y_k, f, h) \qquad k = 0, 1, \ldots. \end{aligned} \tag{12}$$

The method has no memory, so Φ depends only on the arguments listed. We suppose it to be continuous in x and y, and we shall usually not mention $f(x, y)$ and h explicitly. Our treatment of Euler's method had $\Phi(x, y) = f(x, y)$ and we used a Lipschitz condition in an important way. So, for the general procedure we assume that

$$|\Phi(x, u) - \Phi(x, v)| \leq \mathscr{L} \, |u - v| \tag{13}$$

for $a \leq x \leq b$, all $0 < h \leq h_0$ for some h_0, any continuous function f satisfying a Lipschitz condition, and all u, v.

In discussing Euler's method we used as a starting point the fact that the solution $y(x)$ "almost" satisfied the recipe (7). Here we suppose that

$$y(x_{k+1}) = y(x_k) + h\Phi(x_k, y(x_k)) + h\tau_k \tag{14}$$

with τ_k "small." More precisely, if for all x_k in $[a, b]$ and all $h \leq h_0$, there are constants C and p such that

$$|\tau_k| \leq Ch^p,$$

then we shall say that the method is of order p for $y(x)$. This terminology will be justified later, but it is important to note that the order depends on $y(x)$, usually just on how smooth $y(x)$ is. The quantity τ_k is called the local truncation error.

Theorem 2. Suppose we solve the initial value problem

$$y' = f(x, y),$$
$$y(a) = A$$

on the interval $[a, b]$ by the one-step method (12) and suppose that the hypotheses of Theorem 1 are satisfied. If $\Phi(x, y)$ satisfies (13) and if the method is of order $p \geq 1$ for $y(x)$, then for any $x_n = a + nh \in [a, b]$,

$$|y(x_n) - y_n| \leq |y(x_0) - y_0|\, e^{L(x_n - a)} + \frac{Ch^p}{L}(e^{L(x_n-a)} - 1).$$

PROOF. As before, let $e_k = y(x_k) - y_k$ and subtract (12) from (14) to obtain

$$e_{k+1} = e_k + h[\Phi(x_k, y(x_k)) - \Phi(x_k, y_k)] + h\tau_k.$$

Using the Lipschitz condition (13) and the fact that the method is of order p, we see that

$$|e_{k+1}| \leq (1 + h\mathscr{L})\,|e_k| + Ch^{p+1}.$$

The theorem now follows from Lemma 1.

As with our discussion of Euler's method, the result of this theorem gives convergence of $0(h^p)$ provided the initial error is zero (more generally, this error is itself $0(h^p)$). This explains our calling the method of order p for $y(x)$. The term "a method of order p" is used to describe a method which is of order p if f is sufficiently smooth; this gives a false impression if f is not very smooth. Thus, Euler's method is described as a method of order one although it is *not* convergent of order one if $y(x)$ has only one continuous derivative.

The most important task now left is to find functions Φ of order p for smooth $y(x)$ which are inexpensive to evaluate. We need then,

$$y(x_{k+1}) = y(x_k) + h\Phi(x_k, y(x_k)) + h\tau_k$$

with $|\tau_k| \leq Ch^p$. A Taylor series expansion of $y(x)$ shows that

$$y(x_{k+1}) = y(x_k) + h\left[y'(x_k) + \cdots + \frac{h^{p-1}}{p!}y^{(p)}(x_k)\right] + \frac{h^{p+1}}{(p+1)!}y^{(p+1)}(\xi_k)$$

if $y(x) \in C^{p+1}[a, b]$. So we find that if the method is of order p, then it must be the case that

$$\Phi(x, y(x)) = y'(x) + \frac{h}{2!}y''(x) + \cdots + \frac{h^{p-1}}{p!}y^{(p)}(x) + \zeta(x)$$

with $|\zeta(x)| \leq Dh^p$. Hereafter we shall write terms like $\zeta(x)$ as $0(h^p)$. Because $y(x)$ is a solution of the differential equation, $y'(x) = f(x, y(x))$, and we have

$$\begin{aligned}\frac{d}{dx}y'(x) &= \frac{d}{dx}f(x, y(x)) = \frac{\partial}{\partial x}f(x, y(x)) + \frac{\partial}{\partial y}f(x, y(x))\frac{d}{dx}y(x)\\ &= f_x(x, y(x)) + f_y(x, y(x))f(x, y(x)),\end{aligned}$$

where we use subscript notation for partial derivatives. In general, if we use the total derivatives of $f(x, y)$,

$$\begin{aligned}&f^{(1)}(x, y) = f_x(x, y) + f_y(x, y)f(x, y),\\ &f^{(n)}(x, y) = f_x^{(n-1)}(x, y) + f_y^{(n-1)}(x, y)f(x, y) \qquad n = 1, 2, \ldots,\end{aligned}$$

we may easily show that

$$y^{(k)}(x) = f^{(k-1)}(x, y(x)).$$

Thus, for a method of order p we must be able to write

$$\Phi(x, y) = f(x, y) + \frac{h}{2!}f^{(1)}(x, y) + \cdots + \frac{h^{p-1}}{p!}f^{(p-1)}(x, y) + 0(h^p). \quad (15)$$

An obvious choice for Φ then is simply

$$T(x, y) = f(x, y) + \frac{h}{2!}f^{(1)}(x, y) + \cdots + \frac{h^{p-1}}{p!}f^{(p-1)}(x, y).$$

This yields a family of one-step methods called the Taylor series methods. Euler's method is the case $p = 1$. When it is possible to obtain the derivatives analytically, these methods can be very effective, but they are not candidates for general purpose codes.

For a simple equation like that satisfied by Dawson's integral, and especially when high accuracy is desired, the Taylor's series method may be the best way to proceed. This equation has

$$f(x, y) = 1 - 2xy,$$
$$f^{(1)}(x, y) = -2x + (4x^2 - 2)y$$

and so forth. In an exercise we show that a simple recursion formula allows one to use Taylor's series methods of very high order if desired.

The idea of Runge-Kutta methods is to use several evaluations of $f(x, y)$ in a linear combination to approximate $y(x)$. Again, the simplest case is Euler's method. A procedure using two evaluations can be derived as follows. Evaluate $f(x_k, y_k)$; then evaluate $f(\tilde{x}_k, \tilde{y}_k)$ for suitable arguments $\tilde{x}_k, \tilde{y}_k$. Hindsight shows that the algebra is simpler if we write these arbitrary values in the form $\tilde{x}_k = x_k + p_1 h$, $\tilde{y}_k = y_k + p_2 h f(x_k, y_k)$ where now p_1 and p_2 are arbitrary. Then Φ is the linear combination

$$\Phi(x_k, y_k) = a_1 f(x_k, y_k) + a_2 f(\tilde{x}_k, \tilde{y}_k)$$

or better yet,

$$\Phi(x_k, y_k) = a_1 f(x_k, y_k) + a_2 f(x_k + p_1 h, y_k + p_2 h f(x_k, y_k)).$$

In this expression we are free to choose any useful values for p_1, p_2, a_1, a_2. The choice of these parameters is to be made so that the representation (15) holds for as large a value of p as is possible. To carry this out we use Taylor expansions in h and equate coefficients of like powers of h to get as much agreement as possible. To simplify the expressions, we shall not explicitly indicate the arguments of the functions if they are evaluated at (x_k, y_k), but otherwise we shall. It will be necessary to expand a function of two variables in a Taylor's series, since in the general expression the arguments are independent. If the reader is familiar with such expansions, he may skip to the result, but otherwise he can proceed by a succession of familiar one variable expansions as follows:

$$\begin{aligned}
\Phi &= a_1 f + a_2[f(x_k + p_1 h, y_k + p_2 h f)] \\
&= a_1 f + a_2\Big[f(x_k, y_k + p_2 h f) + p_1 h f_x(x_k, y_k + p_2 h f) \\
&\qquad + \frac{p_1^2 h^2}{2} f_{xx}(x_k, y_k + p_2 h f) + 0(h^3)\Big] \\
&= a_1 f + a_2\Big[f + p_2 h f f_y + \frac{p_2^2 h^2}{2} f^2 f_{yy} + 0(h^3) \\
&\qquad + p_1 h f_x + p_1 p_2 h^2 f f_{xy} + 0(h^3) + \frac{p_1^2 h^2}{2} f_{xx} + 0(h^3)\Big] \\
&= (a_1 + a_2) f + a_2 h[p_2 f f_y + p_1 f_x] \\
&\qquad + \frac{a_2 h^2}{2}[p_2^2 f^2 f_{yy} + 2 p_1 p_2 f f_{xy} + p_1^2 f_{xx}] + 0(h^3).
\end{aligned}$$

Now we want to choose the parameters so that

$$\Phi = f + \frac{h}{2}f^{(1)} + \frac{h^2}{6}f^{(2)} + 0(h^3)$$

or, writing this out,

$$\Phi = f + \frac{h}{2}(ff_y + f_x) + \frac{h^2}{6}(f^2 f_{yy} + 2ff_{xy} + f_{xx} + f_x f_y + ff_y^2) + 0(h^3).$$

Equating terms involving the same powers of h, we obtain the requirements

$$\begin{aligned}
h^0: \quad a_1 + a_2 &= 1; \\
h^1: \quad a_2 p_2 &= \tfrac{1}{2}, \\
a_2 p_1 &= \tfrac{1}{2}; \\
h^2: \quad a_2 p_2^2 &= \tfrac{1}{3}, \\
a_2 p_1 p_2 &= \tfrac{1}{3}, \\
a_2 p_1^2 &= \tfrac{1}{3}, \\
? &= f_y f_x/6, \\
? &= ff_y^2/6.
\end{aligned}$$

Evidently it is not possible to obtain agreement in all terms of $0(h^2)$. Let $a_2 = \alpha$. Then for any value of the parameter α,

$$\begin{aligned}
a_2 &= \alpha, \\
a_1 &= 1 - \alpha
\end{aligned}$$

gives a formula with agreement in all terms involving h^0. If we further agree to use only $\alpha \neq 0$, the choice

$$p_1 = p_2 = \frac{1}{2\alpha}$$

gives a formula with agreement in all terms involving h^1. Thus,

$$\Phi = (1 - \alpha)f(x, y) + \alpha f\left(x + \frac{h}{2\alpha}, y + \frac{h}{2\alpha}f(x, y)\right)$$

is a one-step method of order two if $\alpha \neq 0$ and f is sufficiently smooth.

This family of formulas is called the simplified Runge-Kutta family. Special cases have names. Euler's method has $\alpha = 0$ and the order $p = 1$. Heun's method is the case $\alpha = 1/2$, and the modified Euler or Euler-Cauchy method is the case $\alpha = 1$. The broad applicability of these formulas is seen if we ask what restrictions are needed for the convergence theorem to be valid. The continuity of Φ obviously follows from that of f. It is a pleasant fact that the Lipschitz condition on Φ follows from that on f. Since

$$
\begin{aligned}
|\Phi(x, u) - \Phi(x, v)| \\
&= \Bigg|(1 - \alpha)[f(x, u) - f(x, v)] \\
&\quad + \alpha\left[f\left(x + \frac{h}{2\alpha}, u + \frac{h}{2\alpha}f(x, u)\right) - f\left(x + \frac{h}{2\alpha}, v + \frac{h}{2\alpha}f(x, v)\right)\right]\Bigg| \\
&\le |1 - \alpha|\, L\, |u - v| + |\alpha|\, L \left|\left[u + \frac{h}{2\alpha}f(x, u)\right] - \left[v + \frac{h}{2\alpha}f(x, v)\right]\right| \\
&\le |1 - \alpha|\, L\, |u - v| + |\alpha|\, L\, |u - v| + \frac{h}{2} L^2 |u - v| \\
&= \left\{|1 - \alpha| + |\alpha| + \frac{hL}{2}\right\} L\, |u - v|
\end{aligned}
$$

for all $0 < h \le h_0$, we may take the Lipschitz constant for Φ to be

$$
\mathscr{L} = \left\{|1 - \alpha| + |\alpha| + \frac{h_0 L}{2}\right\} L.
$$

Therefore, if the differential equation satisfies the conditions of Theorem 1 and if the function f has two continuous derivatives (which implies the solution $y(x) \in C^3[a, b]$), any of the family of simplified Runge-Kutta formulas with $\alpha \ne 0$ converges of order h^2.

A point of some practical importance arises here. What if we apply a "second-order" Runge-Kutta formula and $y(x)$ belongs to $C^2[a, b]$ but *not* to $C^3[a, b]$? Examination of the local truncation error shows that the formula is then of order one and converges of order h. The point is that the higher order procedures are applicable to less smooth solutions; their rate of convergence is just reduced. In the situation mentioned, Euler's method gets convergence of order h more cheaply, but a code ought to be based on a method of sufficiently high order to take advantage of the average smoothness of solutions presented to it. The odd problem which has slow convergence just makes for a little inefficiency.

Higher order procedures involving more substitutions can be derived in the same way, although naturally the expansions become (very) tedious. As it happens, k evaluations per step lead to procedures of order k for $k = 1, 2, 3, 4$ but *not* for 5. For this reason, fourth order formulas are most common. As in the second order case, there is a family of fourth order procedures depending on several parameters. The "classical" choice of parameters leads to the algorithm:

$$y_0 = A,$$

and for $i = 0, 1, \ldots$

$$\begin{aligned}
k_1 &= f(x_i, y_i), \\
k_2 &= f\left(x_i + \frac{h}{2}, y_i + \frac{h}{2} k_1\right), \\
k_3 &= f\left(x_i + \frac{h}{2}, y_i + \frac{h}{2} k_2\right), \\
k_4 &= f(x_i + h, y_i + hk_3), \\
y_{i+1} &= y_i + \frac{h}{6}(k_1 + 2k_2 + 2k_3 + k_4).
\end{aligned} \tag{16}$$

This is formulated for the first order system of equations

$$\mathbf{Y}(a) = \mathbf{A},$$
$$\mathbf{Y}' = \mathbf{F}(x, \mathbf{Y})$$

in a natural way:

$$\mathbf{Y}_0 = \mathbf{A},$$

and for $i = 0, 1, \ldots$

$$\begin{aligned}
\mathbf{K}_1 &= \mathbf{F}(x_i, \mathbf{Y}_i), \\
\mathbf{K}_2 &= \mathbf{F}\left(x_i + \frac{h}{2}, \mathbf{Y}_i + \frac{h}{2} \mathbf{K}_1\right), \\
\mathbf{K}_3 &= \mathbf{F}\left(x_i + \frac{h}{2}, \mathbf{Y}_i + \frac{h}{2} \mathbf{K}_2\right), \\
\mathbf{K}_4 &= \mathbf{F}(x_i + h, \mathbf{Y}_i + h\mathbf{K}_3), \\
\mathbf{Y}_{i+1} &= \mathbf{Y}_i + \frac{h}{6}(\mathbf{K}_1 + 2\mathbf{K}_2 + 2\mathbf{K}_3 + \mathbf{K}_4).
\end{aligned}$$

A minimum of four function evaluations per step is required for a fourth order Runge-Kutta procedure. The method used in the code RKF requires six evaluations per step. The additional evaluations are necessary if one is to estimate the error being made in a step in order to control it. After discussing errors and their estimation, we give the specific formulas in equations (18) and (19) below.

Errors—Local and Global

Working codes for the initial value problem do not use a fixed step size h. We would like to estimate the error in each step and adjust h accordingly for the next step. A too small value of h means that we are doing unnecessary computation, while a too large value means that we are probably not meeting the desired accuracy requirement. There is an unfortunate confusion on the part of many users of codes with error estimates as to what is being measured and what its relation to the true error is.

The function $y(x)$ denotes the unique solution of the problem

$$y' = f(x, y),$$
$$y(a) = A.$$

By the true or global error at x_{k+1} we mean

$$y(x_{k+1}) - y_{k+1}.$$

Unfortunately, it is relatively difficult and expensive to estimate this quantity. This is perhaps not surprising, since the numerical procedure is only supplied in the previous step with x_k, y_k, and the ability to evaluate f. About the best we can reasonably ask of the numerical procedure is that it come close to following a solution of the differential equation over one step. Let $u(x)$ be the solution of

$$u' = f(x, u),$$
$$u(x_k) = y_k.$$

By the local error at x_k we mean

$$u(x_{k+1}) - y_{k+1},$$

that is, the error we make in one step of following the solution of the differential equation originating at (x_k, y_k). These errors are illustrated in Figure II.10. It is a reasonable request that the numerical procedure keep this error small; what effect this has on the global error depends on the differential equation itself. After all,

$$y(x_{k+1}) - y_{k+1} = \{y(x_{k+1}) - u(x_{k+1})\} + \{u(x_{k+1}) - y_{k+1}\}. \tag{17}$$

The quantity

$$y(x_{k+1}) - u(x_{k+1})$$

is a measure of the "stability" of the differential equation, for if two solutions differ by $y(x_k) - y_k$ at x_k, then they can differ by this quantity at x_{k+1}. If

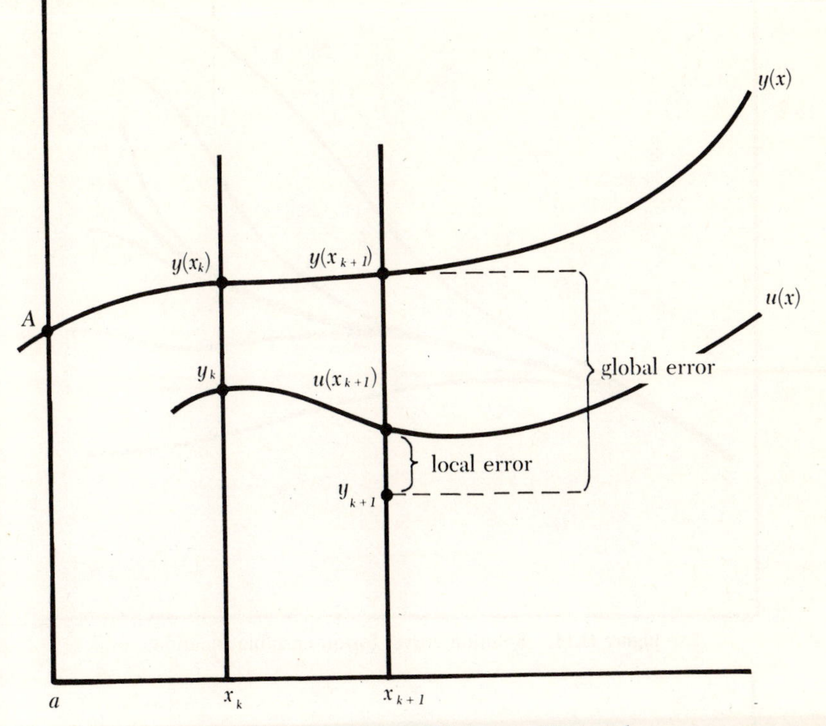

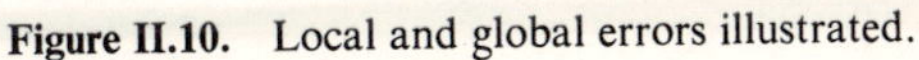

Figure II.10. Local and global errors illustrated.

the quantity increases greatly, we say the problem is poorly posed or ill conditioned or unstable.

By way of a simple example, consider

$$y' = \alpha y$$

for a constant α. Now it is easy to see that

$$y(x) = y(x_k)e^{\alpha(x-x_k)},$$
$$u(x) = y_k e^{\alpha(x-x_k)};$$

so

$$y(x_{k+1}) - u(x_{k+1}) = \{y(x_k) - y_k\}e^{\alpha h}.$$

If $\alpha > 0$, the solution curves spread out (Fig. II.11), the more so as α is large. From the expression (17) it is clear that a small local error at every step does not imply a small global error. On the other hand, if $\alpha < 0$, the curves come together (Fig. II.12) and (17) shows that controlling the local error will control the global error. For general functions $f(x, y)$ the Lipschitz condition alone cannot predict this behavior, since for this example the Lipschitz constant is $|\alpha|$ in either case.

Figure II.11. Solution curves for an unstable equation.

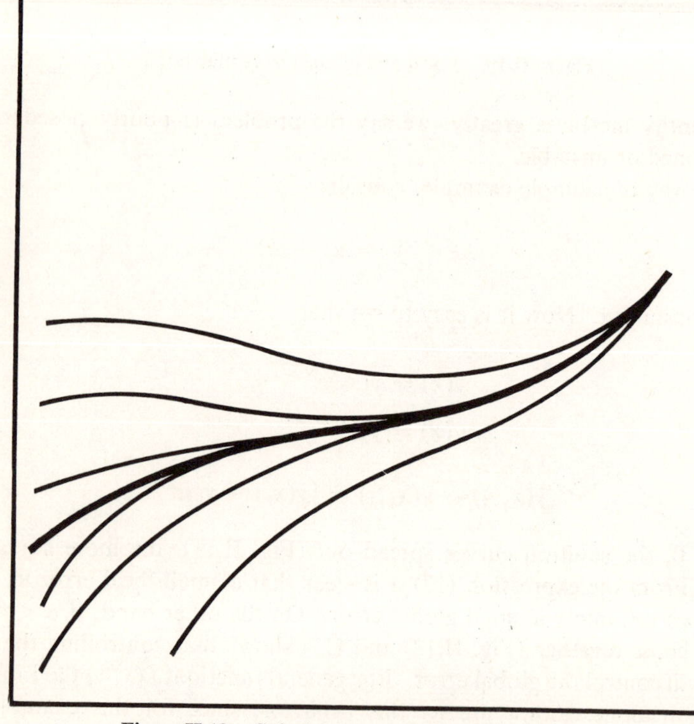

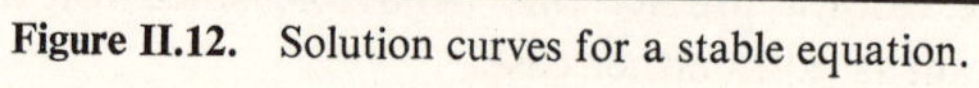

Figure II.12. Solution curves for a stable equation.

Estimating Local Errors

The local error is related to the expression for local truncation error; indeed, it is just h times the local truncation error, μ, for the function $u(x)$:

$$\begin{aligned}\text{local error} &= u(x_{k+1}) - y_{k+1}\\ &= (y_k + h\Phi(x_k, y_k) + h\mu_k) - y_{k+1}\\ &= h\mu_k.\end{aligned}$$

For example, when $y(x)$ is a solution of $y' = f(x, y)$, we have seen that Euler's method has

$$y(x_{k+1}) = y(x_k) + hf(x_k, y(x_k)) + \frac{h^2}{2}(ff_y + f_x) + 0(h^3).$$

Applying this to $u(x)$ shows

$$\text{local error} = h\mu_k = \frac{h^2}{2}(ff_y + f_x) + 0(h^3).$$

Similarly, for the simplified Runge-Kutta formulas of order two ($\alpha \neq 0$), we have

$$u(x_{k+1}) = y_k + h\left[f + \frac{h}{2}f^{(1)} + \frac{h^2}{6}f^{(2)}\right] + 0(h^4),$$

and

$$\tilde{y}_{k+1} = y_k + h\tilde{\Phi}(x_k, y_k)$$

where

$$\tilde{\Phi} = f + \frac{h}{2}(ff_y + f_x) + \frac{h^2}{8\alpha}(f^2f_{yy} + 2ff_{xy} + f_{xx}) + 0(h^3).$$

This leads to

$$\text{local error} = h\tilde{\mu}_k = h^3\left(\frac{1}{6} - \frac{1}{8\alpha}\right)(f^2f_{yy} + 2ff_{xy} + f_{xx}) + \frac{h^3}{6}(f_xf_y + ff_y^2) + 0(h^4).$$

A little reflection about these expressions suggests an interesting possibility. Suppose we are using Euler's method but we also compute a value $\tilde{y}_{k+1}$ by one of these simplified Runge-Kutta formulas. The expressions above show that

$$\begin{aligned}\tilde{y}_{k+1} - y_{k+1} &= \frac{h^2}{2}(ff_y + f_x) + 0(h^3)\\ &= h\mu_k + 0(h^3).\end{aligned}$$

That is, the discrepancy between the two values estimates the error in the lower order formula. This idea is perfectly general. Suppose we are using

$$y_{k+1} = y_k + h\Phi(x_k, y_k)$$

and the truncation error $\mu_k = 0(h^p)$. Suppose we also use

$$\tilde{y}_{k+1} = y_k + h\tilde{\Phi}(x_k, y_k)$$

with truncation error $\tilde{\mu}_k = 0(h^q)$ of a higher order, $q \geq p + 1$. Then by definition

$$\begin{aligned} u(x_{k+1}) &= y_k + h\Phi(x_k, y_k) + h\mu_k \\ &= y_{k+1} + h\mu_k \end{aligned}$$

and similarly

$$u(x_{k+1}) = \tilde{y}_{k+1} + h\tilde{\mu}_k$$

which, on subtracting, shows that

$$\begin{aligned} \tilde{y}_{k+1} - y_{k+1} &= h\mu_k - h\tilde{\mu}_k \\ &= h\mu_k + 0(h^{q+1}). \end{aligned}$$

Because $h\tilde{\mu}_k$ goes to zero faster than $h\mu_k$, we can estimate the local error by

$$\text{local error} = h\mu_k \doteq \tilde{y}_{k+1} - y_{k+1}.$$

For a practical scheme with $p = 4$, the computation of y_{k+1} requires (at least) four function evaluations, and the companion scheme of order $q \geq 5$ requires at least six in general. To be economically feasible, one must try to choose a pair of formulas in which the function evaluations can do double duty. There are a number of parameters at our disposal in generating a fifth order scheme, so that we can attempt to choose them to form an economical pair. Fehlberg [15] has produced a suitable set of formulas:

$$\begin{aligned}
\mathbf{K}_1 &= \mathbf{F}(x_i, \mathbf{Y}_i), \\
\mathbf{K}_2 &= \mathbf{F}\left(x_i + \frac{h}{4}, \mathbf{Y}_i + \frac{h}{4}\mathbf{K}_1\right), \\
\mathbf{K}_3 &= \mathbf{F}\left(x_i + \frac{3h}{8}, \mathbf{Y}_i + h(\tfrac{3}{32}\mathbf{K}_1 + \tfrac{9}{32}\mathbf{K}_2)\right), \\
\mathbf{K}_4 &= \mathbf{F}\left(x_i + \frac{12h}{13}, \mathbf{Y}_i + h(\tfrac{1932}{2197}\mathbf{K}_1 - \tfrac{7200}{2197}\mathbf{K}_2 + \tfrac{7296}{2197}\mathbf{K}_3)\right), \\
\mathbf{K}_5 &= \mathbf{F}(x_i + h, \mathbf{Y}_i + h(\tfrac{439}{216}\mathbf{K}_1 - 8\mathbf{K}_2 + \tfrac{3680}{513}\mathbf{K}_3 - \tfrac{845}{4104}\mathbf{K}_4)), \\
\mathbf{K}_6 &= \mathbf{F}\left(x_i + \frac{h}{2}, \mathbf{Y}_i + h(-\tfrac{8}{27}\mathbf{K}_1 + 2\mathbf{K}_2 - \tfrac{3544}{2565}\mathbf{K}_3 + \tfrac{1859}{4104}\mathbf{K}_4 - \tfrac{11}{40}\mathbf{K}_5)\right).
\end{aligned} \tag{18}$$

Then

$$\mathbf{Y}_{i+1} = \mathbf{Y}_i + h(\tfrac{25}{216}\mathbf{K}_1 + \tfrac{1408}{2565}\mathbf{K}_3 + \tfrac{2197}{4104}\mathbf{K}_4 - \tfrac{1}{5}\mathbf{K}_5) \tag{19a}$$

is the fourth order formula. The fifth order formula is

$$\tilde{\mathbf{Y}}_{i+1} = \mathbf{Y}_i + h(\tfrac{16}{135}\mathbf{K}_1 + \tfrac{6656}{12825}\mathbf{K}_3 + \tfrac{28561}{56430}\mathbf{K}_4 - \tfrac{9}{50}\mathbf{K}_5 + \tfrac{2}{55}\mathbf{K}_6) \tag{19b}$$

which estimates the error in the fourth order formula to be

$$\mathbf{R} = \tilde{\mathbf{Y}}_{i+1} - \mathbf{Y}_{i+1} = h(\tfrac{1}{360}\mathbf{K}_1 - \tfrac{128}{4275}\mathbf{K}_3 - \tfrac{2197}{75240}\mathbf{K}_4 + \tfrac{1}{50}\mathbf{K}_5 + \tfrac{2}{55}\mathbf{K}_6).$$

We only have to make six function evaluations per step to proceed in this way because of the overlap. This means a cost of two additional evaluations per step for the estimate of the local error. The cost is justified because it provides some assurance that the error is being controlled and that the results are reliable. This estimate can also be used to adjust the step size so as to use the largest step consistent with our error requirements. The increased efficiency more than compensates for the cost. The next section discusses this change of step size.

**Step-size Strategy*

We shall devise a strategy for choosing step sizes based on Fehlberg's scheme with its local error estimator. Other possibilities for a strategy may occur to the reader; ours is just one reasonable approach which is developed here in the form of several rules. It is the local error we control, but we hope that the global error will also be controlled. If we keep the local error down to ϵ per unit change in x, meaning that in going from x_i to x_{i+1} we keep

$$|\text{l.e.}| \leq \epsilon(x_{i+1} - x_i),$$

then we do control the global error. A proof of this is very easy, but it requires an elementary result about differential equations which we do not presume the reader knows and which we do not want to develop. A proof of the following theorem can be found in G. Birkhoff and G.-C. Rota, *Ordinary Differential Equations* (Blaisdell, Waltham, Mass., 1969), which is also an excellent general reference book.

Theorem 3. Suppose we solve the initial value problem

$$y' = f(x, y),$$
$$y(a) = A$$

on the interval $[a, b]$ by the one-step method (12) and suppose that the hypotheses of Theorem 1 are satisfied. If $\Phi(x, y)$ satisfies (13) and if the local error is kept down to ϵ per unit step at each step, then

$$|y(x_n) - y_n| \leq \frac{\epsilon}{L}(e^{L(x_n - a)} - 1)$$

for any $x_n \in [a, b]$.

Suppose we have just computed a y_1 from y_0 using Fehlberg's scheme above. Let "est" be the estimate of the local error relative to h, namely

$$\text{est} = \frac{r}{h} = \tfrac{1}{360}k_1 - \tfrac{128}{4275}k_3 - \tfrac{2197}{75240}k_4 + \tfrac{1}{50}k_5 + \tfrac{2}{55}k_6.$$

If we had started at x_0 with a step of length γh, the local error would be

$$\text{l.e.} = (\gamma h)^5 \tau_0 + 0(h^6)$$

for a suitable constant τ_0; so we have

$$\text{l.e.} = \gamma^5(h^5\tau_0) + 0(h^6) = \gamma^5 h \times \text{est} + 0(h^6).$$

If we keep the local error down to ϵ per unit change in x, the allowable error in a step of length γh is $\gamma h \epsilon$. If $h \times |\text{est}|$ is greater than $h \times \epsilon$, we reject the computation of y_1 and recompute with a smaller step length γh, $0 < \gamma < 1$. Since

$$|\text{l.e.}| \doteq \gamma^5 h \times |\text{est}|$$

and we would like $|\text{l.e.}| \leq \gamma h \epsilon$, we have a criterion for choosing γ. For efficiency's sake we would prefer h as large as possible, $|\text{l.e.}| \doteq \gamma h \epsilon$. Then

$$\gamma h \epsilon \doteq \gamma^5 h \times |\text{est}|$$

yields $\gamma = (\epsilon/|\text{est}|)^{1/4}$. As a practical matter, this is too ambitious since we do not want to have many rejections, and aiming at equality with the maximum local error permissible will cause frequent overshoot. So we aim at a fraction of the allowable error or, equivalently, use a fraction of this γ. It is relatively expensive to reject a step with Fehlberg's scheme, so we have been conservative in the code RKF by using a factor of 0.8. This corresponds to aiming at an error of about 0.41ϵ.

If the estimated local error does meet the test, we accept y_1 and continue the computation. Arguing as before, we conclude that it is appropriate to adjust h.

Rule 1. If $|\text{est}| > \epsilon$, reject y_1 and recompute it with a new step size. If $|\text{est}| \leq \epsilon$, accept y_1 and use a new step size for the next step. In either case use

$$0.8(\epsilon/|\text{est}|)^{1/4}h$$

as the new step size.

Some limits on the step size h are necessary. In order to evaluate the quantities $\mathbf{K}_1, \mathbf{K}_2, \ldots$ it is necessary to form expressions like $x + h$ and $x + \frac{12}{13} h$ in floating point arithmetic. To distinguish the various arguments, we do not permit h to become smaller than 13 units of roundoff in x; this is a fundamental limitation imposed by the precision of our arithmetic. A code may reduce its step size to a very small value near a point where the solution has some kind of discontinuity. We feel that in most cases one is not concerned about this, provided the code successfully continues the integration and does not consume too much time. For this reason we have counted function evaluations in our code to provide a measure of computational effort, rather than to request the user to supply a minimum step size, as is common. A maximum step size, however, is quite important. Any code can skip over interesting phenomena unless the step is suitably restricted, so we require the user to provide a maximum step size.

Rule 2. Do not permit the step size to be larger than a user-supplied maximum h_{max}. When at x, do not permit h to be smaller than 13 units of roundoff in x.

Some limits on the change in step size are of practical importance. A very large change can mean that the approximations valid for "small" h are breaking down. Another important consideration is that of "chattering" in the step size. Suppose we are following a very smooth solution which has a discontinuity at an isolated point, and that our working value of h is large and carries us past the discontinuity. After finding a large error present, the code rejects the step and takes a very small one. This step is too small, so the code then wants to take a very large step which puts us past the discontinuity again, and we repeat the cycle.

Rule 3. Do not permit the step size to increase by more than a factor of 5 or decrease by more than a factor of 1/10.

In summary, the user must supply a desired local accuracy ϵ per unit change in x, a nominal starting value of h, and a maximum step size h_{max}. The code estimates the local accuracy and tries to use as large a step as is compatible with the error requirement.

Miscellaneous

It is worthwhile considering whether or not a given problem will yield to analytical techniques, since a general solution may be more informative than a set of tables or graphs. There are books which play the same role for differential equations that integral tables do for integration; they contain techniques for solution and a compendium of equations with known solutions.

Two good books of this kind are G. M. Murphy, *Ordinary Differential Equations and Their Solutions* (Van Nostrand Reinhold Co., New York, 1960) and E. Kamke, *Differentialgleichungen, Lösungsmethoden und Lösungen* (Chelsea, New York, 1948).

In connection with Euler's method, we briefly commented on computational errors, and it is easy to see that the same analysis applies to general one-step methods. If the error in evaluating $\Phi(x_k, y_k)$ is ϵ_k with $|\epsilon_k| \le \epsilon$ and the error in forming $y_k + h[\Phi(x_k, y_k) + \epsilon_k]$ is ρ_k with $|\rho_k| \le \rho$, then

$$\max |y(x_n) - y_n| \le |y(x_0) - y_0| \, e^{L(b-a)} + \frac{(e^{L(b-a)} - 1)}{L}\left(\frac{ch^p}{2} + \epsilon + \frac{\rho}{h}\right).$$

It is the quantity ρ which is the more serious because of the factor $1/h$, but there happens to be a relatively cheap way to reduce its effect. So far we have presumed that all computations were being done in single precision. The function evaluations required in getting $\Phi(x_k, y_k)$ are the bulk of the computational effort, so we do these in single precision. The product of the two single precision numbers h and $[\Phi(x_k, y_k) + \epsilon_k]$ can be represented exactly in double precision. If we form the exact product and add it to y_k in double precision, the effect of roundoff error in the addition is greatly reduced. Most computers form the double precision product automatically when multiplying two single precision numbers. If one can take advantage of this and do the subsequent double precision addition, the only cost has been doing double precision additions instead of single. There is also some extra storage needed for the y_k, which are carried along in double precision and truncated when used in evaluating Φ. Most FORTRANs do not permit the direct addition, and one must perform the multiplication in double precision, too. In any event it is not expensive to use some double precision in this way—it is called partial double precision—and it can be very helpful in controlling the effects of roundoff. The codes supplied have been tested on three major computers, an IBM 360/67, a PDP-10, and a CDC 6600. The long word length of the CDC machine makes this device pointless for the accuracies intended for the code RKF.

To illustrate the effect of roundoff error, we integrated $y(0) = 1$, $y' = -y$ from 0 to 1 using the modified Euler method and a fixed step size of $h = 2^{-n}$ on an IBM 360/67 and a PDP-10. The error at 1 is reported for both single and partial double precision in Table II.3.

Several points are of interest. Arithmetic on the PDP-10 is done in rounded binary, and we see that partial double precision here is not worth the effort. On the other hand, the IBM 360 arithmetic is done in chopped hexadecimal, and the differences between single and partial double precision are striking. The error decreases with decreasing step over this range of h using partial double precision, but with just single precision the accuracy becomes worse for the smallest h instead of better. It is interesting to note that the error of single precision is smaller than that of the partial double

TABLE II.3. EFFECT OF ROUNDOFF ERROR IN SINGLE AND PARTIAL DOUBLE PRECISION.

	Error Single		*Error Partial Double*	
n	IBM 360/67	PDP-10	IBM 360/67	PDP-10
1	2.3(−2)	2.3(−2)	2.3(−2)	2.3(−2)
2	4.6(−3)	4.6(−3)	4.6(−3)	4.6(−3)
3	1.1(−3)	1.1(−3)	1.1(−3)	1.1(−3)
4	2.5(−4)	2.5(−4)	2.5(−4)	2.5(−4)
5	6.1(−5)	6.1(−5)	6.1(−5)	6.1(−5)
6	1.4(−5)	1.5(−5)	1.5(−5)	1.5(−5)
7	1.6(−6)	3.8(−6)	3.8(−6)	3.8(−6)
8	−3.1(−6)	9.6(−7)	9.7(−7)	9.4(−7)
9	−8.2(−6)	2.6(−7)	2.7(−7)	2.3(−7)
10	−1.6(−5)	5.4(−8)	9.4(−8)	5.8(−8)

precision for $h = 2^{-6}$ and 2^{-7}. This is because the effect of the roundoff here is of opposite sign to the effect of the inherent error of the method itself and is cancelling it out.

Most real problems seem to involve functions which are at least piecewise smooth, and if the problems are solved properly, our theorems apply. In some cases coefficients change discontinuously, and in others their derivatives have jumps. An example of a discontinuous change arises when integrating the differential equations describing the motion of a rocket with several stages. The equations depend on the mass of the rocket; at certain times a burnt-out stage is detached and we continue following the payload. The splitting off of a stage means that the mass changes discontinuously at this time. A less dramatic change often occurs when coefficients in the differential equation are obtained from interpolation to physical data. For example, suppose that for

$$y'' + p(x)y' + q(x)y = f(x),$$

the functions p, q, and f are continuous functions made up of cubic polynomials on $[a, z_1]$, $[z_1, z_2]$, $[z_2, b]$ and we wish to integrate y on $[a, b]$. One just has to integrate on the three subintervals separately. This amounts to making sure that the points z_1 and z_2 are in the set $\{x_0, x_1, \ldots, x_N\}$ when integrating from a to b, because with a one-step method the computations are really independent. On all of $[a, b]$, $y(x)$ is only twice continuously differentiable, but on the separate pieces it is infinitely differentiable. As long as we are careful to place mesh points at discontinuities, $y(x)$ "looks" very smooth. Solving such problems will be examined in more detail in the exercises.

To further see how our theorems are more broadly applicable than is apparent, let us consider the example

$$y' = 1 + y^2,$$
$$y(0) = 0.$$

The solution is $y(x) = \tan x$, which becomes infinite at $x = \pi/2$. In general, we consider

$$y' = f(x, y),$$
$$y(a) = A$$

for f with $\partial f/\partial y$ possibly unbounded; in this example $\partial f/\partial y = 2y$. Suppose that we are computing $y(x)$ and that if $|y(x)| \geq M$, we plan to exit with an indication that $y(x)$ is "very" large; perhaps we choose M so that overflow occurs. Suppose further that

$$L = \max_{\substack{a \leq x \leq b \\ -M \leq y \leq M}} \left| \frac{\partial f}{\partial y}(x, y) \right|$$

is finite; in the example, $L = 2M$. Consider the auxiliary problem

$$u' = f^*(x, u),$$
$$u(a) = A$$

where

$$\begin{aligned} f^*(x, u) &= f(x, u) \quad \text{if} \quad |u| \leq M, \\ &= f(x, M) \quad \text{if} \quad u > M, \\ &= f(x, -M) \quad \text{if} \quad u < -M. \end{aligned}$$

The new function is continuous and satisfies a Lipschitz condition with constant L. We leave the proof to the reader. Thus, our theorems apply to this new problem. Now $|A| < M$; so for some interval $[a, c]$, we have $|u(x)| \leq M$. But on this interval

$$u'(x) = f^*(x, u(x)) = f(x, u(x)),$$
$$u(a) = A.$$

Theorem 1 implies that $y(x) \equiv u(x)$ on this interval. Observe that we do not have to make any changes to our code; specifically, we do not actually change $f(x, y)$. Just begin integrating, and if $y(x)$ becomes too large, quit. So the problem for $\tan x$ is not difficult at all; its solution is completely routine except for the fact that our computed solution (correctly) becomes too large at some point short of $\pi/2$ and our code stops.

EXERCISES

1. Verify the statement in the text that for any constant c such that $0 \leq c \leq b$, the function $y(x)$ defined by

$$\begin{aligned} y(x) &= 0, \quad \text{if} \quad 0 \leq x \leq c, \\ &= \tfrac{1}{4}(x - c)^2, \quad \text{if} \quad c < x \leq b \end{aligned}$$

is a solution of the initial value problem

$$y' = \sqrt{|y|},$$
$$y(0) = 0.$$

2. Verify the statement in the text that Dawson's integral is a solution of the initial value problem

$$y' = 1 - 2xy,$$
$$y(0) = 0.$$

3. Consider the problem

$$y' = \sqrt{|1 - y^2|},$$
$$y(0) = 1.$$

Verify that (i) $y(x) \equiv 1$ is a solution on any interval containing $x = 0$.
(ii) $y(x) = \cosh x$ is a solution on $[0, b]$ for any $b > 0$.
(iii) $y(x) = \cos x$ is a solution on a suitable interval. What is the largest interval containing $x = 0$ on which $\cos x$ is a solution?

4. The first order linear differential equation can be solved by integration. If the problem is

$$y' = g(x)y + h(x),$$
$$y(a) = A,$$

show that

$$y(x) = u(x)[A + v(x)]$$

where

$$u(x) = \exp\left[\int_a^x g(t)\,dt\right],$$

$$v(x) = \int_a^x h(t)/u(t)\,dt.$$

You should show that $u(x)$ and $v(x)$ are well defined and have continuous derivatives if the differential equation has continuous coefficients. Would you expect this "explicit" solution to be more useful than numerical solutions if you just want values of $y(x)$? If you want to study the qualitative behavior of $y(x)$? Solve problem 2 by using this general solution. Use the code RKF to integrate the differential equation for Dawson's integral and compare to values in standard tables.

5. Do the following functions satisfy a Lipschitz condition? If so, give suitable constants.
(a) $f(x, y) = 1 + y^2$ for $0 \le x \le \pi/2$,
(b) $f(x, y) = 1 - 2xy$ for $0 \le x \le b$,
(c) $f(x, y) = y/x$ for $1 \le x \le 2$,
(d) $f(x, y) = y/x$ for $-1 \le x \le 1$,
(e) $f(x, y) = \cos x \sin y$ for $-10^6 \le x \le 10^6$.

6. Put the following problems in standard form; differentiation is with respect to t.
   (a) $u^{(4)} + e^t u' - tu = \cos \alpha t$;
   (b) $u'' + v' \cos t + u = t, \quad v' + u' + v = e^{\alpha t}$;
   (c) $u'' + 3v' + 4u + v = 8t, \quad u'' - v' + u + v = \cos t$;
   (d) $mx'' = X(t, x, y, z, x', y', z')$,
   $my'' = Y(t, x, y, z, x', y', z')$,
   $mz'' = Z(t, x, y, z, x', y', z')$;
   (e) $u^{(6)} + uu' = e^t$.

7. Argue that if $f(x, y)$ satisfies a Lipschitz condition, then so does the function $\Phi(x, y)$ for Fehlberg's fourth order Runge-Kutta formula.

8. Consider the linear equation

$$y' = P_1(x)y + Q_1(x).$$

Show that the derivatives needed in the Taylor's series one-step methods can be obtained from

$$y^{(r)} = P_r(x)y + Q_r(x)$$

where

$$P_r(x) = P'_{r-1}(x) + P_1(x)P_{r-1}(x),$$
$$Q_r(x) = Q'_{r-1}(x) + Q_1(x)P_{r-1}(x), \qquad r = 2, 3, \ldots.$$

Apply this by developing a fifth order formula for computing Dawson's integral.

9. An interesting fact about Runge-Kutta methods is that the error depends on the form of the equation as well as on the solution itself. To see an example of this, show that $y(x) = (x + 1)^2$ is the solution of each of the two problems

$$y' = 2(x + 1), \qquad y(0) = 1,$$
$$y' = 2y/(x + 1), \qquad y(0) = 1.$$

Then show that Heun's method is exact for the first equation. Prove that the method is *not* exact when applied to the second equation, although it has the same solution.

10. Consider the initial value problem

$$y''' - y'' \sin x - 2y' \cos x + y \sin x = \ln x,$$
$$y(1) = A_1,$$
$$y'(1) = A_2,$$
$$y''(1) = A_3.$$

Show that the solution $y(x)$ satisfies the first integral relation

$$y''(x) - y'(x)\sin x - y(x)\cos x = c_2 + x \ln x - x$$

and the second integral relation

$$y'(x) - y(x)\sin x = c_1 + c_2 x + \tfrac{1}{2}x^2 \ln x - \tfrac{3}{4}x^2.$$

What are c_1, c_2 in terms of A_1, A_2, A_3? Integrate this problem numerically and monitor the accuracy of your solution by seeing how well it satisfies the integral relations. Argue that if the integral relations are nearly satisfied, then the numerical solution may or may not be accurate, but that if they are not satisfied, the numerical solution must be inaccurate.

11. Convert the code RKF to a partial double precision code. Compare the accuracy of the single precision code to this new code on some simple problems. Is it worthwhile to use partial double precision on your machine?

12. Implement Euler's method and a local error estimator based on Heun's method. Apply it to the problem

$$y' = 10(y - x), \qquad y(0) = 1/10$$

and compare the estimated local error to the true local error. Also compare the global error at several points to the general size of the local errors made in the computations up to this point.

13. Devise a step size strategy for Euler's method with a local error estimator based on Heun's method. Implement it in a code for a single equation. Test it on some of the problems of this section and compare it to the fourth order code provided.

14. An important equation of nonlinear mechanics is van der Pol's equation:

$$x''(t) + \epsilon(x^2(t) - 1)x'(t) + x(t) = 0$$

for $\epsilon > 0$. Regardless of the initial conditions, all solutions of this equation converge to a unique periodic solution which is called a stable limit cycle. For $\epsilon = 1$, choose some initial conditions t_0, $x(t_0)$, and $x'(t_0)$, and integrate the equation numerically until you have apparently converged to the limit cycle. A convenient way to view this is to plot $x'(t)$ against $x(t)$—a phase plane plot. In the phase plane, a periodic solution corresponds to a closed curve. You might like to use Liénard's variables in the integration.

15. Let us consider the problem

$$y' = 2\,|x|\,y, \qquad y(-1) = 1/e$$

on the interval $[-1, 1]$. Verify that the existence and uniqueness theorem (Theorem 1) applies to this problem. Verify that the solution to this problem is

$$y(x) = \begin{cases} e^{x^2} & x \geq 0, \\ e^{-x^2} & x < 0, \end{cases}$$

and further that $y(x)$ has one continuous derivative on $[-1, 1]$ but does not have two. Is Euler's method convergent and $0(h)$ for this problem? What about "higher order" Runge-Kutta methods? What are the answers to these questions for the two problems

$$y' = 2\,|x|\,y, \qquad y(-1) = 1/e$$

on $[-1, 0]$ and

$$y' = 2\,|x|\,y, \qquad y(0) \text{ given}$$

on [0, 1]? Show that if in solving the original problem on $[-1, 1]$, one places a mesh point at $x = 0$, this is the same as solving these two problems. Explain the following numerical results: A fourth order Runge-Kutta code was used to integrate

$$y' = 2\,|x|\,y, \qquad y(-1) = 1/e$$

from $x = -1$ to $x = 1$ using a fixed step size h and the true error at $x = 1$ computed from the analytical solution. Two computations were done. One used $h = 2/2^k$ and the other $h = 2/3^k$. Using an IBM 360/67, the following results were obtained:

k	error, $h = 2/2^k$	error, $h = 2/3^k$
1	$-6.7\,(-3)$	$+3.4\,(-3)$
2	$-5.8\,(-4)$	$+4.5\,(-4)$
3	$-3.1\,(-5)$	$+7.7\,(-5)$
4	$+9.5\,(-6)$	$+5.9\,(-5)$
5	$+2.0\,(-5)$	$+1.6\,(-4)$
6	$+3.7\,(-5)$	$+3.9\,(-4)$

Notice that for the same k, the smaller step size gave *worse* results in almost every case!

16. The Jacobian elliptic functions $sn(x)$, $cn(x)$, $dn(x)$ satisfy the initial value problem

$$y_1' = y_2 y_3, \qquad y_2' = -y_1 y_3, \qquad y_3' = -k^2 y_1 y_2,$$
$$y_1(0) = 0, \qquad y_2(0) = 1, \qquad y_3(0) = 1$$

where k^2 is a parameter between 0 and 1 and $y_1(x) = sn(x)$, $y_2(x) = cn(x)$, $y_3(x) = dn(x)$.

Numerically compute these functions. Check your accuracy by monitoring the relations

$$sn^2(x) + cn^2(x) \equiv 1,$$
$$dn^2(x) + k^2 sn^2(x) \equiv 1,$$
$$dn^2(x) - k^2 cn^2(x) \equiv 1 - k^2.$$

Argue that if these relations are well satisfied numerically, one cannot conclude that the computed functions are accurate but rather that their errors are correlated. If the relations are not satisfied, the functions must be inaccurate. Thus, this test is a necessary test for accuracy but it is not sufficient.

The Jacobian elliptic functions are periodic. You can get the true solutions for $k^2 = 0.51$ from the fact that the period is $4K$ where $K = 1.86264\,08023\,32738\,55203\,02812\,20579\cdots$. If $t_j = jK$, $j = 1, 2, 3, \ldots$, the solutions are given by the relation

$$y_i(t_{j+4}) = y_i(t_j)$$

and the following table:

j	$y_1(t_j)$	$y_2(t_j)$	$y_3(t_j)$
0	0	1	1
1	1	0	0.7
2	0	−1	1
3	−1	0	0.7

17. Show that the solution $Y_1(x)$ of the system

$$Y_1' = -\frac{x}{2} Y_1 + Y_2, \qquad Y_2' = (\tfrac{1}{2} - \tfrac{3}{4}x^2) Y_1 - \frac{x}{2} Y_2$$

satisfies

$$y''(x) + xy'(x) + x^2 y(x) = 0.$$

18. Solve the initial value problem

$$y(0) = 0, \qquad y' = 1 + y^2$$

and compare the computed solution to the true solution $y(x) = \tan x$ for $x = 0(0.1)1.6$. The true solution becomes infinite at $x = \pi/2 \doteq 1.57$. What does your program do? Is the code reasonably effective for this problem?

19. The code RKF normally limits its step size h so that $|h| \leq h_{\max}$, but there is one circumstance in which this is not the case. What is it?

20. The code RKF stretches its step size h by as much as 25 % when it is trying to hit the output point $A + DA = B$. If it did not do this, one might easily encounter IFLAG = 4 unnecessarily or take an inefficiently small step. Why?
   Why is the factor 1.25 used in the test to stretch to B? To answer this, consider two other questions. What is the step size that the code estimates will be successful? What step size does it actually use?

Part III

NUMERICAL LINEAR ALGEBRA

The theory and practice of computation is further advanced in the area of linear algebra than in analysis. This is true for many reasons: The mathematical treatment of linear problems is easier and more complete than for nonlinear problems; digital computers are constructed to do algebraic operations; many important problems can be solved in a finite number of steps, and the like. We give an elementary treatment of the solution of a system of linear equations. The idea of residual correction is not treated for two reasons. First, it is unnecessary for well posed problems, which are the object of this text. Second, the extent and level of the analysis required is incompatible with our intended demands on the reader. To go into this matter, and indeed all of linear systems, in more depth one could turn to the text *Computer Solution of Linear Algebraic Systems* by Forsythe and Moler [19] or a brief treatment by Wilkinson [47]. The definitive treatment at this time is to be found in the two texts of Wilkinson, *Rounding Errors in Algebraic Processes* [45] and *The Algebraic Eigenvalue Problem* [46]. This latter volume is primarily aimed at the problem of computing eigenvalues and eigenvectors, which is an important aspect of numerical linear algebra that we do not treat at all. Lastly, we mention the use of modern methods of orthogonalization such as the modified Gram-Schmidt process (see, for example, Björk [6] and Rice [39]) and Householder transformations (see Businger and Golub [7]) to solve least squares problems. These methods are applicable to much more general problems than the special technique for polynomial fitting developed in Section II.1.2. There is a particularly fine collection of programs made by Wilkinson and Reinsch [48] to solve a great variety of problems in numerical linear algebra.

SYSTEMS OF LINEAR EQUATIONS

Basic Concepts

One of the most basic problems in linear algebra is to solve n simultaneous linear equations in n unknowns $x_1, x_2, \ldots, x_n$:

$$\begin{aligned}
a_{11}x_1 + a_{12}x_2 + \cdots + a_{1n}x_n &= b_1 \\
a_{21}x_1 + a_{22}x_2 + \cdots + a_{2n}x_n &= b_2 \\
\vdots \qquad\qquad\qquad\quad & \;\;\vdots \\
a_{n1}x_1 + a_{n2}x_2 + \cdots + a_{nn}x_n &= b_n.
\end{aligned} \tag{1}$$

The given data here are the right hand sides b_i, $i = 1, 2, \ldots, n$ and the coefficients a_{ij} for $i, j = 1, 2, \ldots, n$. Problems of this nature arise almost everywhere in the applications of mathematics. The student unfamiliar with the origin of such problems can do no better than to consult the books by B. Noble, *Applications of Undergraduate Mathematics in Engineering* [34] and *Applied Linear Algebra* [35]. Each of these books sketches the background and derivation of systems of linear equations in many contexts.

To talk about (1) conveniently we shall use some notation from matrix theory. However, we do not presume that the student has any prior background in this area of mathematics. A student who has encountered one and two dimensional arrays in FORTRAN is ready for all the notation we use.

An $n \times n$ matrix $A = (a_{ij})$ is an array with n rows and n columns of the form

$$A = \begin{bmatrix} a_{11} & a_{12} & \cdots & a_{1n} \\ a_{21} & a_{22} & \cdots & a_{2n} \\ \cdot & & & \\ \cdot & & & \\ \cdot & & & \\ a_{n1} & a_{n2} & \cdots & a_{nn} \end{bmatrix};$$

the element a_{ij} is the entry at the intersection of row i and column j. A vector of n components $\mathbf{b} = (b_i)$ is an array with n rows and 1 column,

$$\mathbf{b} = \begin{bmatrix} b_1 \\ b_2 \\ \cdot \\ \cdot \\ \cdot \\ b_n \end{bmatrix}.$$

The notion of multiplying a matrix A times a vector $\mathbf{x}$ with components $x_1, x_2, \ldots, x_n$ to obtain a vector result $\mathbf{b}$ is defined by equations (1); i.e., the vector $A\mathbf{x}$ has components equal to the left hand sides in (1). In matrix notation (1) is written as

$$A\mathbf{x} = \mathbf{b}.$$

Consider for the moment the case of $n = 1$ in (1),

$$a_{11}x_1 = b_1.$$

If $a_{11} \neq 0$, the equation has a unique solution, namely $x_1 = b_1/a_{11}$. If $a_{11} = 0$, then some problems do not have solutions ($b_1 \neq 0$) while others have many solutions (if $b_1 = 0$, any number x_1 is a solution). In the general situation of $n \geq 1$, some matrices A are such that there is a unique solution vector $x_1, x_2, \ldots, x_n$ for any given right hand side $\mathbf{b}$; other matrices A are such that for some $\mathbf{b}$ there is no solution to (1) at all and for other $\mathbf{b}$ there are many solutions. Every matrix belongs to one of these two classes. Matrices in the second class are called singular matrices, and those in the first class, non-singular. We shall consider only problems with non-singular matrices.

Example. The problem

$$2x_1 + 3x_2 = 8,$$
$$5x_1 + 4x_2 = 13$$

is non-singular; it has the unique solution $x_1 = 1$, $x_2 = 2$. The problem

$$2x_1 + 3x_2 = 4,$$
$$4x_1 + 6x_2 = 7$$

is singular. With this right hand side,

$$\mathbf{b} = \begin{bmatrix} 4 \\ 7 \end{bmatrix},$$

there is no solution; for if x_1 and x_2 were numbers such that $4 = 2x_1 + 3x_2$, then we would have $8 = 2 \times 4 = 2 \times (2x_1 + 3x_2) = 4x_1 + 6x_2$, which is impossible because of the second equation. On the other hand, if

$$\mathbf{b} = \begin{bmatrix} 4 \\ 8 \end{bmatrix},$$

there are many solutions, namely

$$x_1 = \frac{4 - 3c}{2}, \qquad x_2 = c$$

for all real numbers c.

Elimination with Partial Pivoting and Scaling

The most popular method for solving a non-singular problem (1) is called elimination; it is both simple and effective. In principle it can be used to compute solutions of singular problems when they have solutions, but there are better ways to do this; in any case, we are going to avoid singular problems in this text. The basic idea in elimination is to manipulate the equations of (1) in such a way as to arrive at another set of equations that is equivalent to the original set and is easier to solve. By an equivalent set we mean one which has the same solutions. There are two basic operations used in elimination. These are multiplying an equation by a constant, and subtracting a multiple of one equation from another. First, if any equation of (1) is multiplied by the non-zero constant α, we clearly obtain an equivalent set of equations. To see this, suppose that we multiply the kth equation by α to get

$$\alpha a_{k1}x_1 + \alpha a_{k2}x_2 + \cdots + \alpha a_{kn}x_n = \alpha b_k. \tag{2}$$

If $x_1, x_2, \ldots, x_n$ satisfy (1), then they obviously satisfy the set of equations that is the same as (1) except for the kth equation, which is (2). Conversely, because $\alpha \neq 0$, if $x_1, x_2, \ldots, x_n$ satisfy this second set of equations, they obviously satisfy the first. Second, suppose we replace equation i by the result of subtracting the multiple α of equation k from equation i:

$$\begin{aligned}
&a_{11}x_1 + a_{12}x_2 + \cdots + a_{1n}x_n = b_1\\
&\quad\vdots\\
&a_{i-1,1}x_1 + a_{i-1,2}x_2 + \cdots + a_{i-1,n}x_n = b_{i-1}\\
&(a_{i1} - \alpha a_{k1})x_1 + (a_{i2} - \alpha a_{k2})x_2 + \cdots + (a_{in} - \alpha a_{kn})x_n = b_i - \alpha b_k \qquad (3)\\
&\quad\vdots\\
&a_{n1}x_1 + a_{n2}x_2 + \cdots + a_{nn}x_n = b_n.
\end{aligned}$$

If $x_1, x_2, \ldots, x_n$ satisfy (1), then by definition

$$a_{i1}x_1 + a_{i2}x_2 + \cdots + a_{in}x_n = b_i$$

and

$$a_{k1}x_1 + a_{k2}x_2 + \cdots + a_{kn}x_n = b_k$$

so that

$$(a_{i1}x_1 + a_{i2}x_2 + \cdots + a_{in}x_n) - \alpha(a_{k1}x_1 + \cdots a_{kn}x_n) = b_i - \alpha b_k.$$

Thus $x_1, x_2, \ldots, x_n$ satisfy all the equations of (3). To work in reverse, suppose now that $x_1, x_2, \ldots, x_n$ satisfy (3). Then in particular they satisfy equations i and k,

$$(a_{i1} - \alpha a_{k1})x_1 + \cdots + (a_{in} - \alpha a_{kn})x_n = b_i - \alpha b_k$$

and

$$a_{k1}x_1 + \cdots + a_{kn}x_n = b_k,$$

so that

$$[(a_{i1} - \alpha a_{k1})x_1 + \cdots + (a_{in} - \alpha a_{kn})x_n] + \alpha[a_{k1}x_1 + \cdots + a_{kn}x_n] = (b_i - \alpha b_k) + \alpha b_k$$

which is just

$$a_{i1}x_1 + \cdots + a_{in}x_n = b_i.$$

So we see that $x_1, x_2, \ldots, x_n$ satisfy (1).

Let us illustrate elimination by an example. Consider the problem

$$3x_1 + 6x_2 + 9x_3 = 39, \tag{4a}$$

$$2x_1 + 5x_2 - 2x_3 = 3, \tag{4b}$$

$$x_1 + 3x_2 - x_3 = 2. \tag{4c}$$

If we subtract a multiple α of the first equation from the second, we get

$$(2 - 3\alpha)x_1 + (5 - 6\alpha)x_2 + (-2 - 9\alpha)x_3 = 3 - 39\alpha.$$

If we choose $\alpha = 2/3$, the coefficient of x_1 becomes zero and the unknown x_1 no longer appears in this equation:

$$x_2 - 8x_3 = -23.$$

Similarly, we "eliminate" the unknown x_1 from equation (4c) by subtracting 1/3 times the first equation from it. The result of these two operations is the system

$$3x_1 + 6x_2 + 9x_3 = 39, \tag{5a}$$

$$x_2 - 8x_3 = -23, \tag{5b}$$

$$x_2 - 4x_3 = -11. \tag{5c}$$

Let us not concern ourselves with (5a) for the moment, and regard (5b) and (5c) as a pair of equations in the unknowns x_2 and x_3. The same device can be used to eliminate x_2 from (5c). Multiply (5b) by 1 and subtract from (5c) to obtain

$$4x_3 = 12. \tag{6}$$

This is just one equation in the one unknown x_3 and can be solved immediately: $x_3 = \dfrac{12}{4} = 3$. The known value of x_3 can now be used in (5b) to obtain x_2, i.e.,

$$x_2 = 8x_3 - 23 = 8 \times 3 - 23 = 1.$$

The values for x_2 and x_3 can now be used in (5a) to find x_1,

$$\begin{aligned} x_1 &= (-6x_2 - 9x_3 + 39)/3 \\ &= (-6 \times 1 - 9 \times 3 + 39)/3 = 2. \end{aligned}$$

The solution of the problem (4) is then $x_1 = 2$, $x_2 = 1$, $x_3 = 3$.

Let us turn to the general problem (1), which we now write with superscripts to help explain what follows:

$$\begin{aligned} a_{11}^{(1)}x_1 + a_{12}^{(1)}x_2 + \cdots + a_{1n}^{(1)}x_n &= b_1^{(1)}, \\ a_{21}^{(1)}x_1 + a_{22}^{(1)}x_2 + \cdots + a_{2n}^{(1)}x_n &= b_2^{(1)}, \\ \vdots \qquad\qquad & \quad \vdots \\ a_{n1}^{(1)}x_1 + a_{n2}^{(1)}x_2 + \cdots + a_{nn}^{(1)}x_n &= b_n^{(1)}. \end{aligned}$$

If $a_{11}^{(1)} \neq 0$, we can eliminate the unknown x_1 from each of the succeeding equations. A typical step is to subtract from equation i the multiple $a_{i1}^{(1)}/a_{11}^{(1)}$ of the first equation. The results will be denoted with a superscript 2. For $i = 2, 3, \ldots, n$ form the new coefficients

$$a_{ij}^{(2)} = a_{ij}^{(1)} - \frac{a_{i1}^{(1)}}{a_{11}^{(1)}} a_{1j}^{(1)} \qquad j = 1, 2, \ldots, n$$

and

$$b_i^{(2)} = b_i^{(1)} - \frac{a_{i1}^{(1)}}{a_{11}^{(1)}} b_1^{(1)}.$$

Notice that the multiple of the first equation is chosen to make $a_{i1}^{(2)} = 0$, that is, to eliminate the unknown x_1 from equation i. Doing this for each

$i = 2, \ldots, n$ we arrive at the system

$$\begin{aligned}
a_{11}^{(1)}x_1 + a_{12}^{(1)}x_2 + \cdots + a_{1n}^{(1)}x_n &= b_1^{(1)}, \\
a_{22}^{(2)}x_2 + \cdots + a_{2n}^{(2)}x_n &= b_2^{(2)}, \\
a_{32}^{(2)}x_2 + \cdots + a_{3n}^{(2)}x_n &= b_3^{(2)}, \\
&\vdots \\
a_{n2}^{(2)}x_2 + \cdots + a_{nn}^{(2)}x_n &= b_n^{(2)}.
\end{aligned}$$

Now we set the first equation aside and eliminate x_2 from equations $i = 3, \ldots, n$ in the same way. If $a_{22}^{(2)} \neq 0$, then for $i = 3, 4, \ldots, n$ form

$$a_{ij}^{(3)} = a_{ij}^{(2)} - \frac{a_{i2}^{(2)}}{a_{22}^{(2)}} a_{2j}^{(2)} \qquad j = 2, 3, \ldots, n$$

and

$$b_i^{(3)} = b_i^{(2)} - \frac{a_{i2}^{(2)}}{a_{22}^{(2)}} b_2^{(2)}.$$

This results in

$$\begin{aligned}
a_{11}^{(1)}x_1 + a_{12}^{(1)}x_2 + \cdots + a_{1n}^{(1)}x_n &= b_1^{(1)}, \\
a_{22}^{(2)}x_2 + \cdots + a_{2n}^{(2)}x_n &= b_2^{(2)}, \\
a_{33}^{(3)}x_3 + \cdots + a_{3n}^{(3)}x_n &= b_3^{(3)}, \\
&\vdots \\
a_{n3}^{(3)}x_3 + \cdots + a_{nn}^{(3)}x_n &= b_n^{(3)}.
\end{aligned}$$

As before, set the first two equations aside and eliminate x_3 from equations $i = 4, \ldots, n$. This can be done as long as $a_{33}^{(3)} \neq 0$.

The elements $a_{11}^{(1)}, a_{22}^{(2)}, \ldots$ are called pivot elements. Clearly, the process can be continued as long as no pivot vanishes. Assuming this to be the case, we finally arrive at

$$\begin{aligned}
a_{11}^{(1)}x_1 + a_{12}^{(1)}x_2 + \cdots + a_{1n}^{(1)}x_n &= b_1^{(1)}, \\
a_{22}^{(2)}x_2 + \cdots + a_{2n}^{(2)}x_n &= b_2^{(2)}, \\
a_{33}^{(3)}x_3 + \cdots + a_{3n}^{(3)}x_n &= b_3^{(3)}, \\
&\vdots \\
a_{nn}^{(n)}x_n &= b_n^{(n)}.
\end{aligned}$$

This equivalent system is easy to solve by a process known as back substitution. If $a_{nn}^{(n)} \neq 0$, solve for x_n by

$$x_n = b_n^{(n)}/a_{nn}^{(n)}.$$

Using the known value of x_n, solve equation $n - 1$ for x_{n-1} and so forth. Typically, equation k,

$$a_{k,k}^{(k)}x_k + a_{k,k+1}^{(k)}x_{k+1} + \cdots + a_{k,n}^{(k)}x_n = b_k^{(k)},$$

is used to compute x_k from the known values for $x_n, x_{n-1}, \ldots, x_{k+1}$:

$$x_k = \left(b_k^{(k)} - \sum_{j=k+1}^{n} a_{kj}^{(k)}x_j\right)\Big/ a_{kk}^{(k)}.$$

The only way this process can break down (in principle) is if a pivot element is zero. Let us consider some examples:

$$\begin{aligned} 0 \cdot x_1 + 2x_2 &= 3, \\ 4x_1 + 5x_2 &= 6, \end{aligned} \tag{7}$$

$$\begin{aligned} 0 \cdot x_1 + 2x_2 &= 3, \\ 0 \cdot x_1 + 5x_2 &= 6. \end{aligned} \tag{8}$$

In the first example the pivot element $a_{11}^{(1)} = 0$, but there is a simple remedy for the difficulty. We merely interchange the equations, as we realize that the solution of the system clearly does not depend on the order in which the equations are written. So we solve instead the set

$$\begin{aligned} 4x_1 + 5x_2 &= 6, \\ 0 \cdot x_1 + 2x_2 &= 3 \end{aligned}$$

which presents no difficulty. In this case the difficulty was only apparent. This device will not work on the second problem, however, as it is a singular problem. The first equation requires $x_2 = \frac{3}{2}$ while the second requires $x_2 = \frac{6}{5}$; so there is no solution at all.

In the general case, suppose we have arrived at

$$\begin{aligned} a_{11}^{(1)}x_1 + \cdots + a_{1k}^{(1)}x_k + \cdots + a_{1n}^{(1)}x_n &= b_1^{(1)}, \\ \vdots \qquad\qquad \vdots \qquad\qquad \vdots \qquad &\quad \vdots \\ a_{kk}^{(k)}x_k + \cdots + a_{kn}^{(k)}x_n &= b_k^{(k)}, \\ \vdots \qquad\qquad \vdots \qquad &\quad \vdots \\ a_{nk}^{(k)}x_k + \cdots + a_{nn}^{(k)}x_n &= b_n^{(k)} \end{aligned}$$

and the pivot $a_{kk}^{(k)} = 0$. Examine the elements $a_{jk}^{(k)}$ in column k for $j > k$. If for some index ℓ, $a_{\ell k}^{(k)} \neq 0$, interchange equations k and ℓ. Clearly, this does not affect the solution, so let us rename the coefficients in the same way as before; in particular, the new pivot $a_{kk}^{(k)}$ is the old $a_{\ell k}^{(k)} \neq 0$. The elimination process can now proceed as usual. If, however, $a_{jk}^{(k)} = 0$ for all $j = k$, $k + 1, \ldots, n$, we have a difficulty of another sort: the problem is singular. We shall prove this by contradiction. Assume that the problem is non-singular; then $x_1, x_2, \ldots, x_n$ is a solution, and it is the only one. Choose *any* value z_k and set $z_{k+1} = x_{k+1}, \ldots, z_n = x_n$. Then the quantities z_k, $z_{k+1}, \ldots, z_n$ satisfy equations k through n because the unknown x_k does not appear in any of the equations. Now values for $z_1, \ldots, z_{k-1}$ may be determined by back substitution so that equations 1 through $k - 1$ are satisfied:

$$a_{11}^{(1)}x_1 + \cdots + a_{1,k-1}^{(1)}x_{k-1} = b_1^{(1)} - \sum_{i=k}^{n} a_{1i}^{(1)} z_i,$$

$$\vdots$$

$$a_{k-1,k-1}^{(k-1)}x_{k-1} = b_{k-1}^{(k-1)} - \sum_{i=k}^{n} a_{k-1,i}^{(k-1)} z_i.$$

This can be done since none of these pivot elements vanishes. Since all of the equations are satisfied, we have produced a different solution, namely $z_1, z_2, \ldots, z_k, x_{k+1}, \ldots, x_n$ with z_k arbitrary. This contradicts the claim that the system was non-singular.

Example. Here are a couple of examples illustrating how singular systems are revealed to us in the elimination process. In the system

$$\begin{aligned} x_1 + 2x_2 - x_3 &= 2, \\ 2x_1 + 4x_2 + x_3 &= 7, \\ 3x_1 + 6x_2 - 2x_3 &= 7 \end{aligned}$$

one step of elimination yields

$$\begin{aligned} x_1 + 2x_2 - x_3 &= 2, \\ 0 \cdot x_2 + 3x_3 &= 3, \\ 0 \cdot x_2 + x_3 &= 1. \end{aligned}$$

Since we cannot find a non-zero pivot for the second elimination step, the system is singular. The solutions are

$$\begin{aligned} x_1 &= 3 - 2c, \\ x_2 &= c, \\ x_3 &= 1 \end{aligned}$$

for all real numbers c. The system

$$\begin{aligned} x_1 - x_2 + x_3 &= 0, \\ 2x_1 + x_2 - x_3 &= -3, \\ x_1 + 2x_2 - 2x_2 &= -2 \end{aligned}$$

is also singular, since two steps of elimination give

$$\begin{aligned} x_1 - x_2 + x_3 &= 0, \\ 3x_2 - 3x_3 &= 3, \\ 0 \cdot x_3 &= 1. \end{aligned}$$

In this case there is no solution at all.

We conclude that by using interchanges, the elimination process can break down only if the original problem is singular. This argument works in theory, but the distinction between singular and non-singular problems is blurred in practice. Unless a pivot is exactly zero, interchange of equations is unnecessary in theory. However, it is plausible that working with a pivot which is almost zero will lead to problems of accuracy when we are dealing with arithmetic of limited precision. To illustrate this we consider an example of Forsythe and Moler [19]:

$$\begin{aligned} .100 \times 10^{-3}x_1 + 1.00x_2 &= 1.00, \\ 1.00x_1 + 1.00x_2 &= 2.00. \end{aligned} \tag{9}$$

Using three decimal chopped floating point arithmetic, one step in the elimination process without interchanging equations yields

$$\begin{aligned} .100 \times 10^{-3}x_1 + 1.00x_2 &= 1.00, \\ -1.00 \times 10^{4}x_2 &= -1.00 \times 10^{4} \end{aligned}$$

from which it follows that $x_1 = 0.00$, $x_2 = 1.00$. With interchange we have

$$\begin{aligned} 1.00x_1 + 1.00x_2 &= 2.00 \\ 1.00x_2 &= 1.00 \end{aligned}$$

and $x_1 = 1.00$, $x_2 = 1.00$. The true solution rounded to four decimal places is $x_1 = 1.0001$, $x_2 = .9999$. As this example suggests, we must avoid pivots $a_{kk}^{(k)}$ which are small in magnitude. A detailed error analysis of the elimination scheme [45] shows that to control the effect of roundoff errors, the quantities

$$m_{ik} = \frac{a_{ik}^{(k)}}{a_{kk}^{(k)}}$$

must be kept small. In the example with no interchange, $a_{21}^{(1)}/a_{11}^{(1)} = 1.00 \times 10^4$. The element $a_{22}^{(2)}$ is obtained from the quantity

$$1.00 - (1.00 \times 10^4)(1.00).$$

The second term is so large that the term $a_{22}^{(1)} = 1.00$ is ignored; i.e., the large value of $a_{21}^{(1)}/a_{11}^{(1)}$ causes the information in $a_{22}^{(1)}$ to be lost.

Choosing as a pivot at the ith step the largest in magnitude of the quantities $a_{ik}^{(k)}$, $i = k, k + 1, \ldots, n$ is a simple way to guarantee $|m_{ik}| \leq 1$. This strategy of choosing the largest element in a column as the pivot is called a partial pivoting strategy. The code LASSOL does elimination with partial pivoting as described above. In the code, rows are physically interchanged. On most computers it is more efficient to keep track of the interchanges in an auxiliary array, but a code based on physical interchanges rather than index manipulations is easier to read and understand.

The "size" of the elements in the various equations must be comparable for the partial pivoting strategy to be useful. The process of scaling equations (multiplying each equation through by some constant) may be helpful in this regard. Scaling has no direct effect on the answers, but it does affect the solution process through the choice of pivot elements. It can have an adverse effect, as the following example shows. The problem (9) behaved badly if we choose the first element $a_{11}^{(1)}$ as pivot. The partial pivoting strategy chose $a_{21}^{(1)}$ as pivot with good results. However, suppose we multiply the first equation of (9) by 10^5 and the second by 1 to obtain

$$10.0x_1 + 1.00 \times 10^5 x_2 = 1.00 \times 10^5,$$

$$1.00x_1 + 1.00x_2 = 2.00.$$

Since $10.0 > 1.00$, the pivoting strategy chooses 10.0 as pivot, and elimination yields

$$10.0x_1 + 1.00 \times 10^5 x_2 = 1.00 \times 10^5,$$

$$- 1.00 \times 10^4 x_2 = -1.00 \times 10^4.$$

The solution using three digit chopped arithmetic is $x_1 = 0.00$ and $x_2 = 1.00$, which is poor (as we might expect because this is the same answer we get without pivoting in the original system).

A reasonable method of scaling is to divide each equation by its largest element in magnitude; i.e., divide equation i by

$$\max_{1 \leq j \leq n} |a_{ij}|,$$

which results in the largest coefficient in equation i being 1 in magnitude. When this process of scaling is used in (1), we say that the associated matrix A has been "row equilibrated."

In very difficult problems, the roundoff errors arising in the scaling process can actually make results with the scaled matrix worse. It is possible to avoid doing the scaling explicitly and thus eliminate this source of error entirely. The technique is explained by Forsythe and Moler in *Computer Solution of Linear Algebraic Equations* [19]. Alternatively, one can do scaling using only powers of the base of the machine's floating point arithmetic, since the divisions can then be done without rounding error. (This amounts to merely shifting the decimal point.) For simplicity and clarity, we have scaled explicitly in LASSOL as described above. It is uncommon for this detail to matter much except in pathological problems, and this text is aimed toward the solution of the more frequently occurring problems rather than the unusual.

Accuracy

Ordinarily, elimination with partial pivoting as applied to a row equilibrated matrix gives answers that are accurate in a certain sense. We shall give our attention to a way of assessing the accuracy of computed solutions rather than to the theory for predicting accuracy. Perhaps the most natural test of accuracy is to measure how well the computed solution satisfies the original equations. If the given problem is $A\mathbf{x} = \mathbf{b}$ and we compute $\mathbf{z}$, the residual vector $\mathbf{r}$ is defined to be $\mathbf{r} = \mathbf{b} - A\mathbf{z}$ or in component form

$$r_i = b_i - \sum_{j=1}^{n} a_{ij} z_j, \qquad i = 1, 2, \ldots, n.$$

Since the equations have been scaled, a "large" residual r_i means that we have failed to satisfy equation i; this is clearly a signal for alarm. We should turn to higher precision, a more sophisticated algorithm, or a reformulation of the problem.

It is significant that elimination with partial pivoting is virtually certain to yield a solution $\mathbf{z}$ with small residuals. To get some feeling for this, suppose the matrix A has the special form we call triangular, i.e.,

$$\begin{bmatrix} a_{11}, & a_{12}, & a_{13}, & \ldots, & a_{1n} \\ 0 & a_{22}, & a_{23}, & \ldots, & a_{2n} \\ 0 & 0 & a_{33}, & \ldots, & a_{3n} \\ \cdot & \cdot & & \cdot & \cdot \\ \cdot & \cdot & & \cdot & \cdot \\ \cdot & \cdot & & \cdot & \cdot \\ 0 & 0 & & \cdots & a_{nn} \end{bmatrix}$$

Solving $A\mathbf{x} = \mathbf{b}$ then amounts to just the back substitution. But the back substitution process is nothing more than computing each x_i so as to make

$r_i = 0$. After all,

$$r_i = b_i - a_{ii}x_i - \sum_{j=i+1}^{n} a_{ij}x_j$$

and we compute x_i by

$$x_i = \left(b_i - \sum_{j=i+1}^{n} a_{ij}x_j\right) \Big/ a_{ii}.$$

These special systems are always solved with residuals as small as possible in the arithmetic precision being used. The elimination process applied to a general matrix transforms it into an equivalent triangular matrix which is solved by back substitution. Because of the numerical errors in forming the triangular matrix, we do not end up with an exactly equivalent system. This fact leads to the presence of non-zero residuals, but it is plausible and usually true that they are small.

Suppose that all the residuals are small; is the solution "accurate"? The answer depends on what is desired of a solution and on the problem itself. In many contexts, satisfying all equations with only a small discrepancy is a perfectly adequate definition of "accurate." To a person conscious of the mathematical theory, it is perhaps obvious that one ought to compare the computed solution $\mathbf{z}$ to the $\mathbf{x}$ uniquely defined as the solution of the non-singular problem $A\mathbf{x} = \mathbf{b}$. Computationally, the matter is more subtle and we feel that the residuals are a very practical approach.

Let us explore the relationship between the residual $\mathbf{r}$ and the error $\mathbf{e} = \mathbf{x} - \mathbf{z}$. Now $\mathbf{r} = \mathbf{b} - A\mathbf{z} = A\mathbf{x} - A\mathbf{z} = A(\mathbf{x} - \mathbf{z}) = A\mathbf{e}$ or in component form

$$r_i = \sum_{j=1}^{n} a_{ij}e_j, \qquad i = 1, 2, \ldots, n.$$

This implies

$$|r_i| \leq \sum_{j=1}^{n} |a_{ij}|\,|e_j| \leq n \max |e_j|$$

since A was scaled so that each $|a_{ij}| \leq 1$. Because $\max|r_i| \leq n \max|e_j|$, small errors guarantee small residuals.

The converse is not true; one can have large errors along with small residuals. To see this we argue as follows: If the matrix A is non-singular, the theory asserts that there is a matrix called the inverse of A, denoted by $A^{-1} = (\alpha_{ij})$, such that for any vector $\mathbf{c}$ the solution of $A\mathbf{y} = \mathbf{c}$ is

$$\mathbf{y} = A^{-1}\mathbf{c}.$$

In component form,

$$y_i = \sum_{j=1}^{n} \alpha_{ij}c_j, \qquad i = 1, 2, \ldots, n.$$

(At the end of this chapter we shall show how the code LASSOL can be used to construct A^{-1} if one should want it.) Now since $A\mathbf{e} = \mathbf{r}$, we have

$$e_i = \sum_{j=1}^{n} \alpha_{ij} r_j, \qquad i = 1, 2, \ldots, n. \tag{10}$$

Unfortunately, the scaling of A does not necessarily keep the elements of A^{-1} scaled. It can and does happen that some elements of A^{-1} are large. So we see from (10) that small residuals may not guarantee small errors in the solution components. A simple example is

$$A = \begin{bmatrix} 1, & -1 \\ 1, & -1 + 10^{-5} \end{bmatrix}$$

which has the inverse

$$A^{-1} = \begin{bmatrix} 1 - 10^5, & 10^5 \\ -10^5, & 10^5 \end{bmatrix}.$$

(We ask the reader to verify this in an exercise.) If we were solving $A\mathbf{x} = \mathbf{0}$ so that $\mathbf{x} = \mathbf{0}$ and somehow computed

$$\mathbf{z} = \begin{bmatrix} 1 \\ 1 \end{bmatrix},$$

we would find

$$\mathbf{r} = \begin{bmatrix} 0 \\ 10^{-5} \end{bmatrix}.$$

Both residuals are small, but there is a large error in each solution component.

If we take a more physically oriented view of error, the residual is rehabilitated. By a "well posed" or "well conditioned" problem we mean that if the data are changed by a small amount, then the results are changed by only a small amount. If, for the problem at hand, the right hand side $\mathbf{b}$ and the solution vector $\mathbf{x}$ consist of physically meaningful quantities, one may well be able to claim on physical grounds that the problem is well posed. Then, since $A\mathbf{z} = \mathbf{b} + \mathbf{r}$ and $A\mathbf{x} = \mathbf{b}$, we see that a small $\mathbf{r}$ means $\mathbf{z}$ is close to $\mathbf{x}$.

Another interpretation also argues on physical grounds but is independent of "well posedness." Suppose the entries b_i are measured quantities. Then we do not know the true value c_i but rather a measured value b_i subject to uncertainty and (perhaps) some idea as to its accuracy. For simplicity's sake, suppose we say we know only that $|c_i - b_i| \leq \epsilon_i$. Now $\mathbf{z}$ is the exact solution of $A\mathbf{z} = \mathbf{c} + \mathbf{r}$. If it were the case that $|r_i| \leq \epsilon_i$ for each i, then we could say that $\mathbf{z}$ is the exact solution of $A\mathbf{z} = \mathbf{d}$ (where $d_i = c_i + r_i$) and d_i is as close to the true value c_i as is b_i. With the assumptions made, we have to regard **z** as "exact" since it is the exact solution of a problem that the experimenter cannot distinguish from the one given. This last point of view is more pervasive than one might suspect. It points out the real difficulty with the narrowly mathematical view of error. One speaks of comparing the solution **x** of $A\mathbf{x} = \mathbf{b}$ to the computed solution **z**; the precision of this statement is illusory. A more precise statement requires one to distinguish between the problem he wants to solve, $M\mathbf{y} = \mathbf{c}$, and the problem he actually passes to his code, $A\mathbf{x} = \mathbf{b}$. After all, the computer uses an arithmetic with a finite number of digits and usually a number system other than decimal. Only in exceptional cases can the elements of M and **c** be represented exactly as machine numbers, so in fact we have machine approximations A and **b**. We can expect our code to approximate **x** but not the solution of the problem we have in mind. If the residuals are small and the problem well posed, then **z** is close to **x** which is close to **y**. If the problem is pathologically ill conditioned, it is not unusual that **z** is closer to **x** than **x** is to **y**. That is, we make more error by putting the problem into the machine than we do in solving it in the machine!

At the level of this text we would have to invert the matrix to pinpoint ill conditioning. This is too expensive for general use and is contrary to our intention of considering only well posed problems. The size of the residuals is always an important indicator of accuracy, and for well posed problems one need go no further. However, the reader should appreciate that for badly posed problems one may not be able to take the residuals at face value.

The question now is whether we can compute a reasonably accurate value for **r**. This cannot ordinarily be done in single precision since, roughly speaking, the very nature of elimination tries to make

$$\sum_{j=1}^{n} a_{ij} z_j$$

as close to b_i as is possible in single precision. This means that ordinarily all significant figures of single precision are lost in the subtraction which yields r_i. If one forms the products $a_{ij}z_j$ in double precision and the sum in double precision, he can greatly reduce the error. (Indeed, for most machines the product $a_{ij}z_j$ of two single precision numbers is exactly represented in double precision.) With the rare exception of cases of terms of greatly differing magnitude, the sum is computed correct to several digits more than single precision. If the subtraction of the sum from b_i is done in double precision, there is considerable cancellation but almost invariably one has a residual with a few good leading figures, which are all we require. The reader interested in the details of the analysis and more about systems of linear equations can consult the text of Forsythe and Moler. The code LASSOL makes available with each solution the quantity

$$R = \max_i |r_i|\,,$$

which is computed in double precision.

As was previously mentioned, LASSOL can be used to find the elements of A^{-1}. Although not needed for our purposes in solving (1), it is sometimes needed for other purposes, e.g., in some statistical computations where it is an estimate of certain statistical parameters. Let $\mathbf{x}^i$ denote the solution of

$$A\mathbf{x} = \mathbf{b}^i \quad \text{for} \quad i = 1, 2, \ldots, n$$

where the ith right hand side $\mathbf{b}^i$ is

$$b_j^i = \begin{cases} 0 & \text{if} \quad i \neq j \\ 1 & \text{if} \quad i = j. \end{cases}$$

If we form the matrix

$$X = \begin{bmatrix} x_1^1 & x_1^2 & \cdots & x_1^n \\ x_2^1 & x_2^2 & \cdots & x_2^n \\ \cdot & \cdot & & \cdot \\ \cdot & \cdot & & \cdot \\ \cdot & \cdot & & \cdot \\ x_n^1 & x_n^2 & \cdots & x_n^n \end{bmatrix},$$

it is easy to show that $X = A^{-1}$; we leave this as an exercise.

EXERCISES

In each of the following problems use the code LASSOL where applicable unless specifically stated otherwise.

1. Put the linear system

$$\begin{aligned} -3x_1 + 8x_2 + 5x_3 &= 6, \\ 2x_1 - 7x_2 + 4x_3 &= 9, \\ x_1 + 9x_2 - 6x_3 &= 1 \end{aligned}$$

in the matrix form

$$A\mathbf{x} = \mathbf{b}.$$

Scale the system and then solve by elimination with partial pivoting as described in the text. Do your calculations by hand using exact arithmetic and as a check, form the residuals

$$r_i = b_i - \sum_{j=1}^{3} a_{ij}x_j, \qquad i = 1, 2, 3.$$

2. Repeat problem 1 for the system

$$\begin{aligned} x_2 + x_3 &= 0, \\ x_1 - 5x_2 + 3x_2 &= 0, \\ 2x_1 + x_2 - 4x_3 &= -1. \end{aligned}$$

3. Consider again the system in problem 1. Solve by hand using 3 digit rounded floating point arithmetic (after each arithmetic operation round to three digits) by:

   (a) elimination without partial pivoting or scaling;
   (b) elimination with partial pivoting and no scaling;
   (c) elimination with partial pivoting and scaling.

   Compare with your previous results.

4. Recall that all systems of the form (1) do not possess unique solutions. For example,

$$x_1 + x_2 = 1,$$
$$x_1 + x_2 = 2$$

has no solution at all, while

$$x_1 + x_2 = 2,$$
$$2x_1 + 2x_2 = 4$$

has an infinity of solutions, namely $x_1 = c$, $x_2 = 2 - c$ for all real numbers c. This has an interesting geometrical interpretation. In the plane the problem of determining the solution to

$$a_{11}x_1 + a_{12}x_2 = b_1,$$
$$a_{21}x_1 + a_{22}x_2 = b_2$$

is equivalent to finding the point of intersection of the two lines $a_{11}x_1 + a_{12}x_2 = b_1$ and $a_{21}x_1 + a_{22}x_2 = b_2$. In the first example the lines are parallel, and in the second they coincide. Similar interpretations may be made in higher dimensions.

Using elimination with partial pivoting, determine which of the following systems are singular and which are non-singular. For the non-singular problems, find solutions. Use exact arithmetic.

a) $$\begin{aligned} x_1 + 2x_2 - x_3 &= 2, \\ 2x_1 + 4x_2 + x_3 &= 7, \\ 3x_1 + 6x_2 - 2x_3 &= 7 \end{aligned}$$

b) $$\begin{aligned} x_1 - x_2 + x_3 &= 0, \\ 2x_1 + x_2 - x_3 &= -3, \\ x_1 + 2x_2 - 2x_3 &= -2 \end{aligned}$$

c) $$\begin{aligned} x_1 + x_2 + x_3 &= 0, \\ 2x_1 + x_2 - x_3 &= -3, \\ 2x_1 \qquad - 4x_3 &= -6 \end{aligned}$$

d) $$\begin{aligned} 2x_1 - 3x_2 + 2x_3 + 5x_4 &= 3, \\ x_1 - x_2 + x_3 + 2x_4 &= 1, \\ 3x_1 + 2x_2 + 2x_3 + x_4 &= 0, \\ x_1 + x_2 - 3x_3 - x_4 &= 0 \end{aligned}$$

5. Using the code LASSOL, solve the systems in problems 1 and 2. Compare with your results calculated by hand.

6. Suppose we are given the electrical network shown in Figure III.1 and desire to find the potentials at junctions (1) through (6). The potential applied between A and B is V volts. Denoting the potentials by $v_1, v_2, \ldots, v_6$, application of Ohm's Law and Kirchoff's Current Law yield the following set of linear

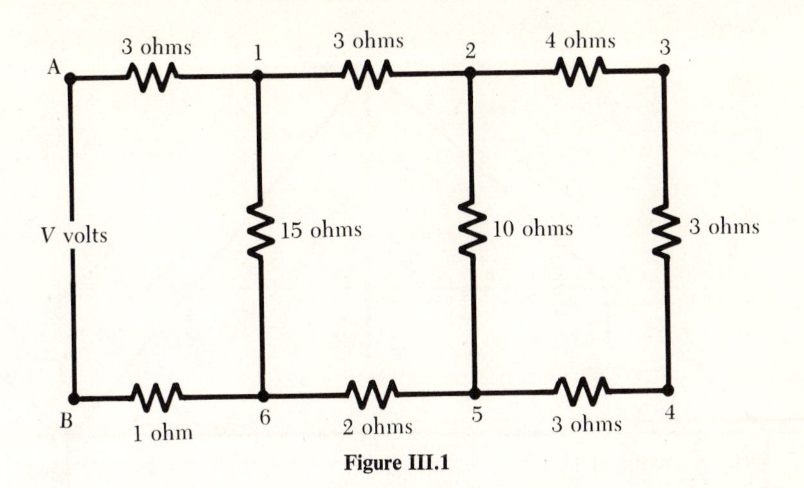

Figure III.1

equations for the v_i:

$$\begin{aligned}
11v_1 - 5v_2 - v_6 &= 5V,\\
-20v_1 + 41v_2 - 15v_3 - 6v_5 &= 0,\\
-3v_2 + 7v_3 - 4v_4 &= 0,\\
-v_3 + 2v_4 - v_5 &= 0,\\
-3v_2 - 10v_4 + 28v_5 - 15v_6 &= 0,\\
-2v_1 - 15v_5 + 47v_6 &= 0.
\end{aligned}$$

Solve for $V = 50, 100, 150$.

7. Consider the A-frame illustrated in Figure III.2. Assuming the base is smooth and neglecting the weight of the members, find

   a) the floor reactions at A and E (R_A, R_E):
   b) the pin reactions at B on AC (B_h, B_v);
   c) the pin reactions at C on CE (C_h, C_v);
   d) the pin reactions at D on BD (D_h, D_v).

   Summing forces and moments, the appropriate equations are

$$\begin{aligned}
8.00\,R_E - 1784.0\,0 &= 0.00,\\
-8.00\,R_A + 1416.00 &= 0.00,\\
C_h + D_h &= 0.00,\\
C_v + D_v + 223.00 &= 0.00,\\
\frac{-5.18\,C_v}{\sqrt{2}} - \frac{5.18\,C_h}{\sqrt{2}} + \frac{446.00}{\sqrt{2}} &= 0.00,\\
-5.77\,D_v - 1456.00 &= 0.00,\\
-5.77\,B_v - 852.00 &= 0.00,\\
B_h + D_h &= 0.00.
\end{aligned}$$

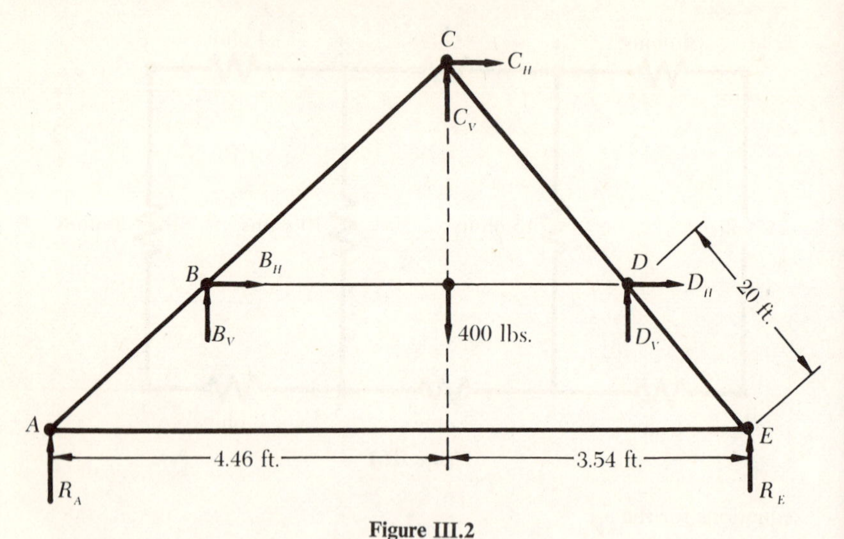

Figure III.2

Solve using the code LASSOL. A little observation will show you that these equations can be solved analytically quite easily. Do so and compare with your numerical results.

8. A 6 ft. by 6 ft. table is mounted on three legs. Four loads are applied as indicated in Figure III.3. Determine the three reactions R_1, R_2, R_3. Summing forces in the z direction, and moments about the x and y axes, leads to the equations

$$R_1 + R_2 + R_3 = 110.00,$$
$$R_1 + R_2 = 78.33,$$
$$R_2 + R_3 = 58.33.$$

9. Many problems of mechanics, mathematical physics, and technology lead to consideration of equations of the form

$$f(x) = g(x) + \lambda \int_a^b K(x, y) f(y)\, dy,$$

where $f(x)$ is an unknown function. These equations are called integral equations, as the unknown function appears under the integral sign. The function $g(x)$ is called the free term, $K(x, y)$ the kernel, and λ the parameter of the equation. We shall assume that the functions g and K are continuous. There are several methods for obtaining approximate solutions to this problem (in very simple cases one can find the solution analytically), but probably the most intuitive is to regard it as a generalized matrix problem. Divide the interval $a \leq y \leq b$ into n equal subintervals by letting $h = \dfrac{b-a}{n}$ and defining

$$x_j = a + jh, \qquad j = 1, \ldots, n.$$

Approximate the integral

$$\int_a^b K(x, y) f(y)\, dy$$

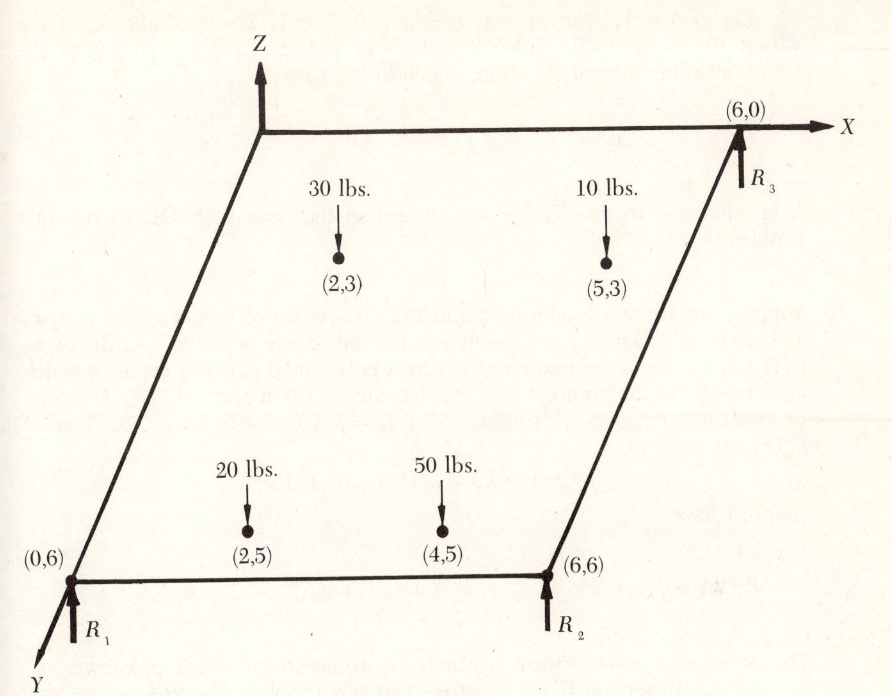

Figure III.3

by the Riemann sum

$$h \sum_{j=1}^{n} K(x, x_j) f(x_j).$$

In the approximate equation

$$f(x) \doteq g(x) + \lambda h \sum_{j=1}^{n} K(x, x_j) f(x_j),$$

replace x by $x_1, \ldots, x_n$ to obtain the linear algebraic system

$$f(x_i) \doteq g(x_i) + \lambda h \sum_{j=1}^{n} K(x_i, x_j) f(x_j), \qquad i = 1, \ldots, n,$$

or

$$f(x_i)[1 - \lambda h K(x_i, x_i)] - \lambda h \sum_{\substack{j=1 \\ i \neq j}}^{n} K(x_i, x_j) f(x_j) \doteq g(x_i), \qquad i = 1, \ldots, n,$$

where the unknown quantities are the $f(x_i)$. If we replace the approximate equality by an equality and solve the resulting system of equations, the solution will yield an approximation to the solution of the integral equation at $x_1, \ldots, x_n$. There will be values of the parameter λ for which the algebraic problem is singular; we avoid these values.

Let $g(x) = 1$, $K(x, y) = x + y$, $a = 0$, $b = 1$, $\lambda = 2$. Take for n the values 10, 20, and 40. Solve the algebraic problem and compare with the solution to the integral equation, which in this case is

$$f(x) = -\tfrac{2}{3}x.$$

You will have to modify the dimensions in the code LASSOL to run this problem.

10. Suppose we know a function $f(x)$ at the discrete set of points $x_0, x_1, \ldots, x_n$ and we want to know f at a point y not equal to one of the x_i. As discussed in II.1.1, we could approximate $f(x)$ by a polynomial $P_n(x)$ of degree n which agrees with f at the points x_i and use the value $P_n(y)$ in place of $f(y)$. One way to construct $P_n(x)$ is as follows. Let $f_i = f(x_i)$, $i = 0, 1, \ldots, n$; then if $P_n(x)$ has the form

$$P_n(x) = a_0 + a_1x^1 + \cdots + a_nx^n,$$

we must have

$$P_n(x_i) = f(x_i) = f_i = a_0 + a_1x_i^1 + \cdots + a_nx_i^n, \qquad i = 0, 1, \ldots, n.$$

This is a set of $n + 1$ linear algebraic equations in the $n + 1$ unknowns a_0, $a_1, \ldots, a_n$. In section II.1.1 we prove that this problem always has one and only one solution, so the matrix must be non-singular.

(a) Let $f(x) = e^x$ be defined at the points $x_i = \dfrac{i}{5}$, $i = 0, 1, \ldots, 5$. Find $P_5(x)$ and evaluate at $x = .12, .23, .45, .67, .89$. Compare with e^x.

(b) This method of computing $P_n(x)$ can be ill conditioned. The divided difference form presented in II.1.1 is in general much simpler and more stable. Let us consider an example. The Cauchy formula

$$n = A + \frac{B}{\lambda^2} + \frac{C}{\lambda^4}$$

is an empirical formula for representing the index of refraction n in terms of the wave length λ. It not only works well for ordinary optical materials in the visible range, but can also be justified on theoretical grounds. Some data for light crown glass are:

λ Wavelength (Angstroms)	n
4000.0	1.5238
4500.0	1.5180
5000.0	1.5139

Using the above method, interpolate to find A, B, and C. What is the maximum residual R? Put the problem in matrix form and find the inverse matrix. Does the problem appear to be well conditioned?

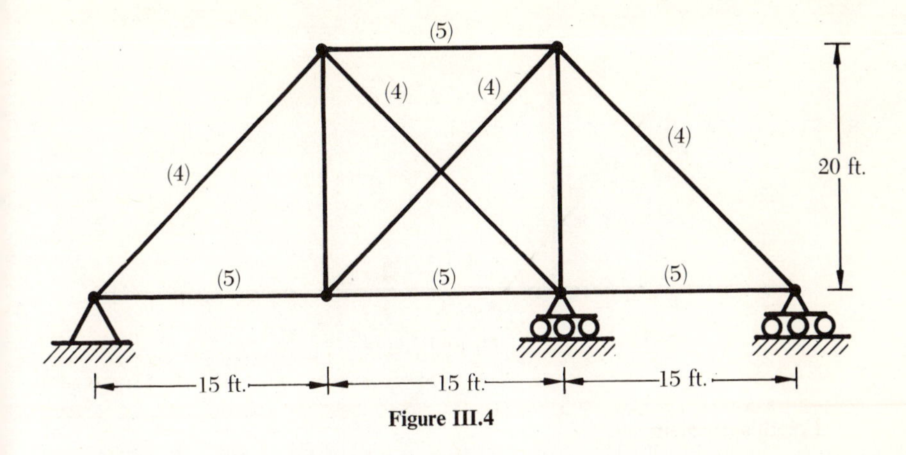

Figure III.4

11. Wang, in *Matrix Methods of Structural Analysis* (International Textbook Company, Scranton, Pa., 1966), considers the statically indeterminate pin-jointed truss shown in Figure III.4, where numbers parenthesized are cross sectional areas of the individual members and $E = 30{,}000$ ksi, a uniform modulus of elasticity. With the problem we associate

    (a) a statics matrix A which defines the configuration of the framework;
    (b) a member stiffness matrix S which relates the elastic properties of the constituent members; and
    (c) an external force vector $\mathbf{p}$ which describes the applied forces at the joints.

    We seek a displacement vector $\mathbf{x}$ (Fig. III.5) which accounts for the displacement at each degree of freedom and an internal force vector $\mathbf{f}$ (Fig. III.6) acting on each member. Using the displacement method of solution, we find that

$$K\mathbf{x} = \mathbf{p}, \qquad K = ASA^T,$$

$$\mathbf{f} = (SA^T)\mathbf{x}.$$

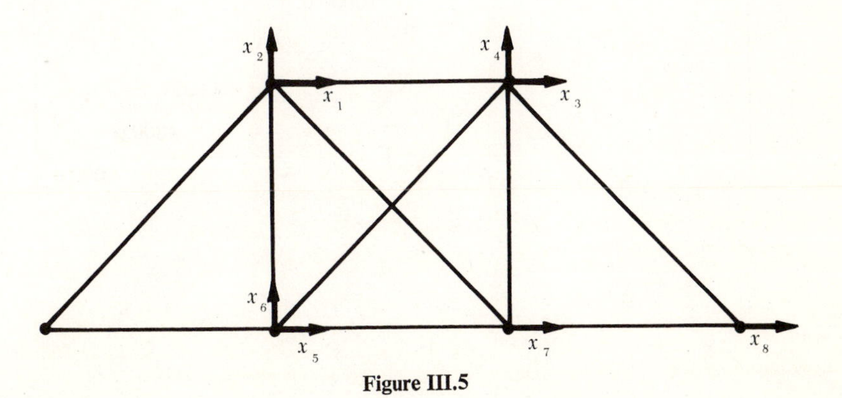

Figure III.5

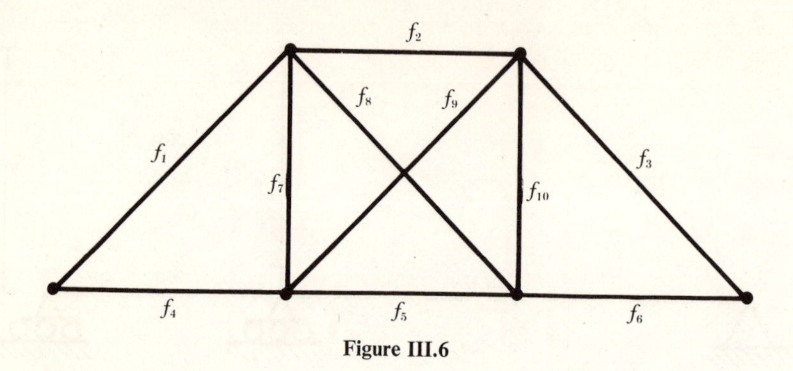

Figure III.6

For this problem

$$A = \begin{bmatrix} 0.6 & -1.0 & 0.0 & 0.0 & 0.0 & 0.0 & 0.0 & -0.6 & 0.0 & 0.0 \\ 0.8 & 0.0 & 0.0 & 0.0 & 0.0 & 0.0 & 1.0 & 0.8 & 0.0 & 0.0 \\ 0.0 & 1.0 & -0.6 & 0.0 & 0.0 & 0.0 & 0.0 & 0.0 & 0.6 & 0.0 \\ 0.0 & 0.0 & 0.8 & 0.0 & 0.0 & 0.0 & 0.0 & 0.0 & 0.8 & 1.0 \\ 0.0 & 0.0 & 0.0 & 1.0 & -1.0 & 0.0 & 0.0 & 0.0 & -0.6 & 0.0 \\ 0.0 & 0.0 & 0.0 & 0.0 & 0.0 & 0.0 & -1.0 & 0.0 & -0.8 & 0.0 \\ 0.0 & 0.0 & 0.0 & 0.0 & 1.0 & -1.0 & 0.0 & 0.6 & 0.0 & 0.0 \\ 0.0 & 0.0 & 0.6 & 0.0 & 0.0 & 1.0 & 0.0 & 0.0 & 0.0 & 0.0 \end{bmatrix},$$

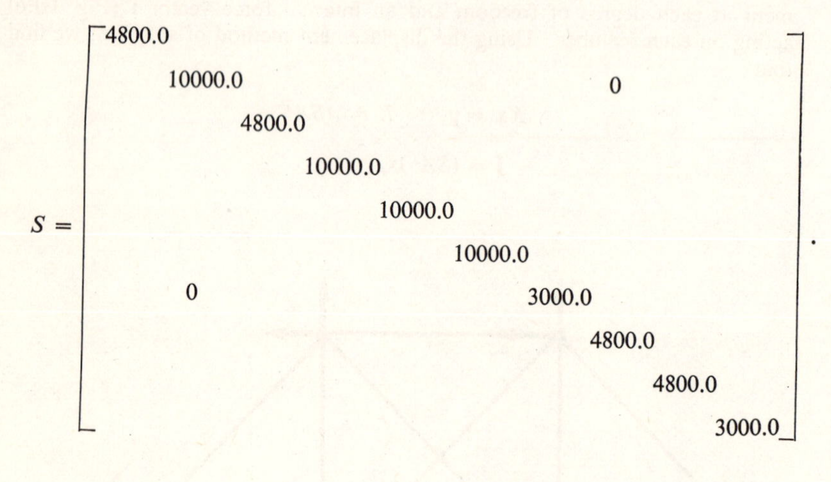

$$S = \begin{bmatrix} 4800.0 & & & & & & & & & \\ & 10000.0 & & & & & & 0 & & \\ & & 4800.0 & & & & & & & \\ & & & 10000.0 & & & & & & \\ & & & & 10000.0 & & & & & \\ & & & & & 10000.0 & & & & \\ & 0 & & & & & 3000.0 & & & \\ & & & & & & & 4800.0 & & \\ & & & & & & & & 4800.0 & \\ & & & & & & & & & 3000.0 \end{bmatrix}.$$

Write a program to form matrix products and determine the elements of K. Solve for $\mathbf{x}$, using the three $\mathbf{p}$ vectors

$$\mathbf{p} = \begin{bmatrix} 0.0 \\ -1.0 \\ 0.0 \\ 0.0 \\ 0.0 \\ 0.0 \\ 0.0 \\ 0.0 \end{bmatrix}, \quad \begin{bmatrix} 0.0 \\ 0.0 \\ 0.0 \\ -1.0 \\ 0.0 \\ 0.0 \\ 0.0 \\ 0.0 \end{bmatrix}, \quad \begin{bmatrix} 0.0 \\ 0.0 \\ 0.0 \\ 0.0 \\ 0.0 \\ -1.0 \\ 0.0 \\ 0.0 \end{bmatrix}.$$

Find the corresponding vectors $\mathbf{f}$.

12. Consider the system

$$\begin{aligned} -3x_1 + 8x_2 + 5x_3 &= 6, \\ -2x_1 - 7x_2 + 4x_3 &= 9, \\ x_1 + 9x_2 - 6x_3 &= 1. \end{aligned}$$

Put the problem in the matrix form $A\mathbf{x} = \mathbf{b}$ and solve using LASSOL. Again using the code, find A^{-1}. As you will recall, the solution to our problem is also given by $A^{-1}\mathbf{b}$. Form $A^{-1}\mathbf{b}$ and compare its residual vector with that of the first solution. Which method of solution requires less computational effort?

13. Consider the problem

$$A\mathbf{x} = \mathbf{b}$$

and define n right hand sides $\mathbf{b}^i$ by

$$b_j^i = \begin{cases} 0 & \text{if } i \neq j \\ 1 & \text{if } i = j, \quad i = 1, 2, \ldots, n. \end{cases}$$

Denote the corresponding solutions by $\mathbf{x}^i$. Verify that the matrix

$$X = \begin{bmatrix} x_1^1 & x_1^2 & \cdots & x_1^n \\ x_2^1 & x_2^2 & \cdots & x_2^n \\ \cdot & \cdot & & \cdot \\ \cdot & \cdot & & \cdot \\ \cdot & \cdot & & \cdot \\ x_n^1 & x_n^2 & \cdots & x_n^n \end{bmatrix}$$

is the inverse of the matrix A, i.e., $X = A^{-1}$.

14. Verify the statement in the text that

$$\begin{bmatrix} 1 - 10^5, & 10^5 \\ -10^5, & 10^5 \end{bmatrix}$$

is the inverse of

$$\begin{bmatrix} 1, & -1 \\ 1, & -1 + 10^{-5} \end{bmatrix}.$$

15. We have observed that if a computed solution **z** of $A\mathbf{x} = \mathbf{b}$ yields a large residual, it must be inaccurate. Another simple test for numerical difficulties considers the size of components of **z**. Suppose the components of A and **b** are of "reasonable" size but **z** has "large" components. Then either **z** is quite inaccurate or the problem is poorly posed (or both). To prove this, consider two cases:
    (i) **x** has no "large" components,
    (ii) **x** has "large" components (Hint: represent **b** in terms of A^{-1}).

16. A matrix A with entries a_{ij} is called "banded" if $a_{ij} = 0$ when $|i - j| > m$ for some m. The band width is $2m + 1$. The code LASSOL is more effective for such matrices if the statement

```
IF(QUOT.EQ.0.0)GO TO 8
```

is inserted just after the statement defining QUOT. Why? What about triangular matrices?

Part IV

SOLUTIONS TO EXERCISES

Most of the numerical solutions given were computed on an IBM 360/67 computer in single precision arithmetic. Some of the computations (e.g., residuals and estimated errors) are very sensitive to word length. If the reader obtains results significantly different from those given, he should consider whether this is due to machine precision or to the organization of the computation. Results given to more than seven digits were computed either in double precision on an IBM 360/67 or on a machine with a longer word length.

SOLUTIONS FOR I

1. In our idealized floating point arithmetic

$$x \otimes y = \text{fl}(x \times y) = \text{fl}(y \times x) = y \otimes x$$

and

$$x \oplus y = \text{fl}(x + y) = \text{fl}(y + x) = y \oplus x.$$

2. i) $x = .801 \times 10^1$, $y = .125 \times 10^1$, $z = .808 \times 10^2$
   ii) $x = .113 \times 10^3$, $y = -.111 \times 10^3$, $z = .751 \times 10^1$
   iii) $x = .200 \times 10^3$, $y = -.600 \times 10^2$, $z = .603 \times 10^1$
   iv) $x = .100 \times 10^1$, $y = .500 \times 10^{-2}$, $z = -.100 \times 10^1$.

3. The discussion in the text developed the expression

$$(x \otimes y) \otimes z = x \otimes (y \otimes z)\left[\frac{(1 + \delta_1)(1 + \delta_2)}{(1 + \delta_3)(1 + \delta_4)}\right]$$

where $|\delta_i| \leq u$ for each $i = 1, 2, 3, 4$. The relative error ϵ in the associative law is $(x \otimes y) \otimes z = x \otimes (y \otimes z)(1 + \epsilon)$. If ϵ is non-negative, then

$$1 + \epsilon = 1 + |\epsilon| = \frac{(1 + \delta_1)(1 + \delta_2)}{(1 + \delta_3)(1 + \delta_4)} \leq \frac{(1 + u)^2}{(1 - u)^2};$$

so

$$|\epsilon| \leq \frac{(1 + u)^2}{(1 - u)^2} - 1 = \frac{4u}{(1 - u)^2}$$

If ϵ is non-positive, then

$$1 + \epsilon = 1 - |\epsilon| \geq \frac{(1 - u)^2}{(1 + u)^2}$$

and

$$|\epsilon| \leq 1 - \frac{(1 - u)^2}{(1 + u)^2} = \frac{4u}{(1 + u)^2} < \frac{4u}{(1 - u)^2}.$$

In any event we see that

$$|\epsilon| \leq \frac{4u}{(1 - u)^2}.$$

4. The approximate solution $z = .180 \times 10^2$ of the problem

$$.111x = .200 \times 10^1$$

has a residual r of

$$.200 \times 10^1 \ominus (.111 \otimes .180 \times 10^2) = .200 \times 10^1 \ominus .199 \times 10^1$$
$$= .100 \times 10^{-1}$$

in the arithmetic of problem 2.

Double precision (6 digit chopped decimal arithmetic) leads to

$$.200000 \times 10^1 \ominus (.111000 \otimes .180000 \times 10^2) =$$
$$.200000 \times 10^1 \ominus .199800 \times 10^1 = .200000 \times 10^{-2}.$$

Since the product az is representable exactly in double precision, exact arithmetic yields the same residual as that of double precision. Obviously the higher precision was needed to get a reasonable approximation to the residual.

5. Assuming that the routine SQRT is exact,

$$\frac{\sqrt{x + x\epsilon} - \sqrt{x}}{\sqrt{x}} \doteq \frac{\epsilon x(\sqrt{x})'}{\sqrt{x}} = \epsilon x \cdot \frac{1}{2\sqrt{x}} \cdot \frac{1}{\sqrt{x}} = \frac{\epsilon}{2}.$$

This says that the relative error in $\sqrt{x}$ is about half that in the approximation to x itself.

If the routine SQRT is such that $\text{SQRT}(y) = \sqrt{y}(1 + \delta)$, then we have the computed approximation $\sqrt{x + x\epsilon}\,(1 + \delta)$ so

$$\frac{\sqrt{x + x\epsilon}\,(1 + \delta) - \sqrt{x}}{\sqrt{x}} = \sqrt{1 + \epsilon}\,(1 + \delta) - 1.$$

Now thinking of ϵ as a variable and using Taylor's expansion about $\epsilon = 0$ shows that

$$\sqrt{1 + \epsilon} \doteq 1 + \frac{\epsilon}{2} \quad \text{for small } \epsilon$$

so

$$\frac{\sqrt{x + x\epsilon}\,(1 + \delta) - \sqrt{x}}{\sqrt{x}} \doteq \frac{\epsilon}{2} + \delta.$$

Consequently, the approximation to $\sqrt{x}$ is still accurate in the sense of relative error.

6. $u = .100 \times 10^{-60}$, $v = .100 \times 10^{60}$, $w = .100 \times 10^{60}$.

7. Not always; take $u = v = .900 \times 10^1$ in the arithmetic of problem 2.

8. This problem was solved on an IBM 360/67, a PDP-10, and a CDC 6600 with the results as indicated below.

Computer	number of terms required	absolute error	relative error
IBM 360/67	45	3.98×10^{-5}	8.76×10^{-1}
PDP-10	48	4.02×10^{-6}	8.86×10^{-2}
CDC 6600	56	3.29×10^{-12}	7.24×10^{-8}

The discrepancy in results is, of course, due to the differences in computer arithmetic, especially the number of digits carried in the single precision calculations. The terms in the sum s_n alternate in sign, increase rapidly in magnitude at first (for $n = 9$, $s_n \doteq -1.413143 \times 10^3$), and then decrease. The actual value of e^{-10} correct to seven figures is 4.539993×10^{-5}. For the partial sums to decrease to this size there is necessarily cancellation. The errors in the first few sums become prominent in the final answer.

A little reflection will show that the series does not exhibit this problem if $x \geq 0$ or if $x < 0$ and "small." The larger $x < 0$ is in magnitude the worse the problem becomes. FORTRAN uses more efficient and accurate methods than this series for computing the exponential.

SOLUTIONS FOR II.1.1

1. No, it is not. For example, let $f(x) = \dfrac{1}{1 + x^2}$ be defined on the nodes $\{0, 1, 2, 3\}$. From the divided difference table:

x	$f(x)$			
0	1			
1	$\frac{1}{2}$	$-\frac{1}{2}$		
2	$\frac{1}{5}$	$-\frac{2}{5}$	$\frac{1}{10}$	
3	$\frac{1}{10}$	$-\frac{3}{10}$	$\frac{1}{10}$	0

we see that

$$\begin{aligned} P_3(x) &= 1 - \tfrac{1}{2}x + \tfrac{1}{10}x(x-1) + 0x(x-1)(x-2) \\ &= 1 - \tfrac{3}{5}x + \tfrac{1}{10}x^2. \end{aligned}$$

In general we have no guarantee that $a_n \neq 0$.

2. Suppose $f(x)$ is a polynomial of degree n or less and let $P_n(x)$ be the polynomial interpolating f on the nodes $x_0, x_1, \ldots, x_n$. As $P_n(x)$ is unique and f is also a polynomial interpolating on the nodes, $P_n(x) \equiv f(x)$. The

result also follows from (6a),

$$P_n(x) = f(x) + R_n(x),$$

and (10),

$$R_n(x) = \frac{f^{(n+1)}(z)\prod_{i=0}^{n}(x - x_i)}{(n+1)!},$$

since $f^{(n+1)}(z) \equiv 0$ for all z implies $R_n(x) \equiv 0$.

As an example, let $f(x) = x^2$ be defined on the nodes $\{-2, -1, 0, 1, 2\}$. By forming the difference table

−2	4				
−1	1	−3			
0	0	−2	1		
1	1	−1	1	0	
2	4	0	1	0	0

we find

$$\begin{aligned} P_4(x) &= 4 - 3(x+2) + 1(x+2)(x+1) \\ &\quad + 0(x+2)(x+1)x + 0(x+2)(x+1)x(x-1) \\ &= x^2. \end{aligned}$$

3. $P_1(x) = 2x$. Examples are $Q(x) = x^2 - x + 2$, $S(x) = -x^3 + 2x^2 + 3x - 2$. This does not contradict our theory, which asserts in this case that there is one and only one polynomial *of degree at most one* which interpolates the data. This polynomial is $P_1(x)$.

4. Consider the case $n = 1$. Using the nodes 3.92 and 3.94 we find that $P_1(3.947) = 1.77496$ while for the nodes 3.94 and 3.96, $P_1(3.947) = 1.78215$. Using one additional node (3.96 in the first case and 3.92 in the second), the error estimates were .00484 and −.00236, respectively. To evaluate the bound we note that

$$f^{(2)}(t) = e^t \quad \text{for all } t.$$

The function e^t increases for $t \geq 0$, so in the first case

$$|f^{(2)}(t)| \leq e^{3.94} = 51.41860 \quad \text{for} \quad 3.92 \leq t \leq 3.94$$

and in the second

$$|f^{(2)}(t)| \leq e^{3.96} = 52.45733 \quad \text{for} \quad 3.94 \leq t \leq 3.96.$$

The corresponding bounds are then .00486 and .00239. We summarize all of this in the following table.

Nodes used	$P_1(3.947)$	Approximate error	Error bound	Actual error
3.92, 3.94	1.77496	0.00481	0.00486	0.00481
3.94, 3.96	1.78215	−0.00236	0.00239	−0.00238

We note that in both cases the approximate error expressions were quite accurate. We would expect interpolation on the nodes 3.94 and 3.96 to be the more accurate of the two (as is the case) because the point $x = 3.947$ is in the interval spanned by 3.94 and 3.96.

5. The results from INTPOL were

Degree n	$P_n(0.14)$	Estimated error	True error
1	0.1568894	1.0×10^{-4}	5.7×10^{-5}
2	0.1569555	-8.6×10^{-6}	-8.4×10^{-6}
3	0.1569468	1.1×10^{-7}	1.7×10^{-7}

To bound the worst error for linear interpolation between the nodes x_i and x_{i+1}, we evaluate the bound

$$\max_{x_i \le t \le x_{i+1}} |f^{(2)}(t)| \cdot \frac{(x_{i+1} - x_i)^2}{8}.$$

Now

$$f'(t) = (\operatorname{erf}(t))' = \frac{2}{\sqrt{\pi}} e^{-t^2}.$$

$$f''(t) = -\frac{4t}{\sqrt{\pi}} e^{-t^2}.$$

The function $|f''(t)|$ is strictly increasing on $0.0 \le t \le 0.2$ since

$$\left| \frac{d}{dt} f''(t) \right| = \left| \frac{4}{\sqrt{\pi}} e^{-t^2}(2t^2 - 1) \right| \ne 0 \quad \text{for} \quad 0 \le t \le \frac{\sqrt{2}}{2} \doteq 0.7.$$

So

$$|f''(t)| < \frac{4(0.20)e^{-(0.20)^2}}{1.772454}$$

for $0.00 \le t \le 0.20$ and our bound becomes

$$\frac{4 \cdot (0.20)e^{-(0.20)^2}}{1.772454} \cdot \frac{(0.05)^2}{8} \doteq 0.000137.$$

The true error in the linear interpolation above was .0000576.

6. If one writes out the product

$$L_k(x) = \prod_{\substack{m=0 \\ m \neq k}}^{n} \frac{x - x_m}{x_k - x_m}$$

$$= \frac{x - x_0}{x_k - x_0} \cdot \frac{x - x_1}{x_k - x_1} \cdot \dots \cdot \frac{x - x_{k-1}}{x_k - x_{k-1}} \frac{x - x_{k+1}}{x_k - x_{k+1}} \cdot \dots \cdot \frac{x - x_n}{x_k - x_n},$$

it is easy to see that

(a) $L_k(x)$ is a product of $n + 1 - 1 = n$ linear factors and hence is a polynomial of degree n;

(b) if $x = x_k$, then all the factors have the value 1 and hence the product has the value 1. If $x = x_m$, $m \neq k$, the factor containing $x - x_m$ is zero and the product vanishes.

(c) if $x = x_i$, then $P_n(x_i) = \sum_{k=0}^{n} L_k(x_i) f_k = L_i(x_i) f_i = f_i$.

When one interpolation is to be carried out, the amounts of computation required by the divided difference form and the Lagrange form are roughly equivalent. Less computer storage is required for the Lagrange form, since there is no need to store the divided difference table.

The Lagrange form gives prominence to the function values $f(x_k)$ and the divided difference form to the nodes x_k. If one wants to interpolate many functions at the same place, the Lagrange form is preferable. If one wants to vary the nodes, the divided difference form is preferable. Because our estimate of the error is based on raising the degree of the interpolating polynomial or, equivalently, adding a node, the divided difference form is ordinarily preferred in practical computation. However, the Lagrangian form is usually better suited to the theoretical analysis of interpolation and its applications.

7. Let x_i, x_{i+1}, x_{i+2} be three consecutive nodes. The quadratic polynomial $P_2(x)$ interpolating $f(x)$ on these nodes has the form

$$P_2(x) = f_i + f_{i,i+1}(x - x_i) + f_{i,i+1,i+2}(x - x_i)(x - x_{i+1}).$$

Differentiating, we find

$$P_2''(x) = 2f_{i,i+1,i+2}.$$

We shall use P'' to approximate f'' at $x = x_{i+1}$. From the sketch in Figure IV.1, we see that the equivalence point is near 23.1. Approximating the

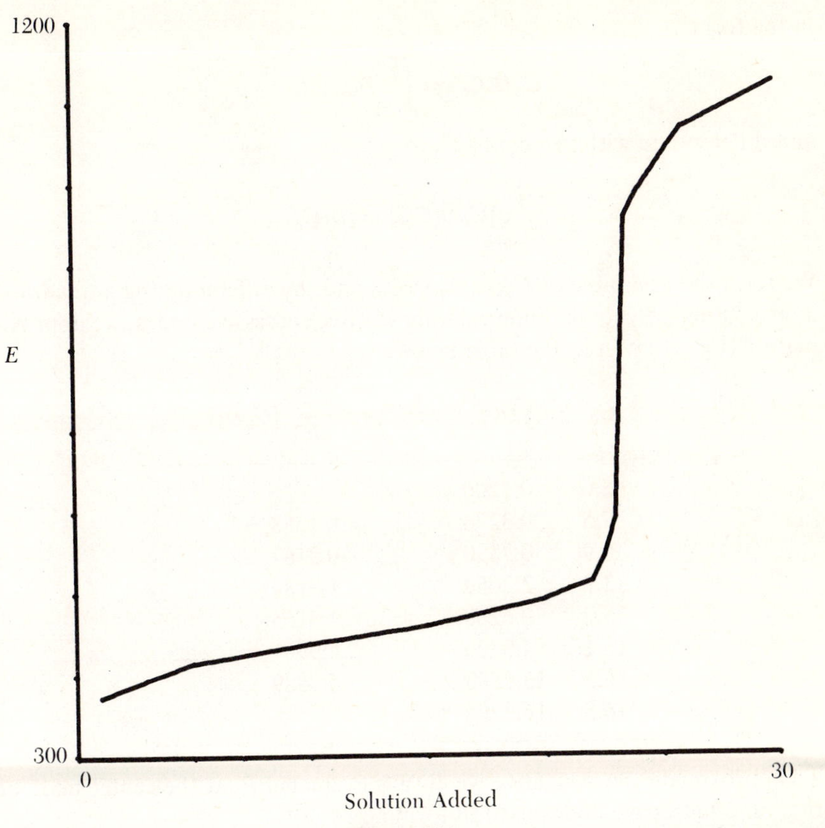

Figure IV.1

second derivative by P_2'' in the manner described above, we have the following values:

Solution Added	E	P_2''
22.90	575.00	
23.00	590.00	1500
23.10	620.00	21000
23.20	860.00	−18500
23.30	915.00	−2600
23.40	944.00	

The equivalence point is apparently between $x = 23.1$ and $x = 23.2$.

8. To obtain an expression for $D(C_0)$, we write the equation

$$\bar{D}(C_0) = \frac{1}{C_0} \int_0^{C_0} D(C)\, dc$$

in the form

$$C_0\bar{D}(C_0) = \int_0^{C_0} D(C)\,dc$$

and differentiate with respect to C_0 to get

$$\frac{d}{dC_0}[C_0\bar{D}(C_0)] = D(C_0)$$

We form the new table of C_0 vs. $C_0\bar{D}(C_0)$ and, by differentiating a quadratic interpolating polynomial interpolating at three successive nodes, we approximate $D(C_0)$ as given in the table below.

C_0	$C_0\bar{D}(C_0)$	$D(C_0) = \frac{d}{dC_0}[C_0\bar{D}(C_0)]$
5.0	0.1200	
7.5	0.3278	0.1388
9.9	0.7890	0.3167
12.9	2.2059	1.3184
13.2	2.6268	1.3765
15.1	4.9226	4.9967
16.3	13.7898	5.7589
16.8	16.3296	

To approximate the derivative we always evaluated P_2' at the center node of the three successive nodes used in forming it.

9. Suppose there are $n + 1$ distinct nodes and the polynomial $P_m(x)$ of degree $m < n$ satisfies $P_m(x_k) = f(x_k)$ for all k. The result developed in the text says that the interpolation conditions $P_m(x_k) = f(x_k)$ for $k = 0, 1, \ldots, m$ already *uniquely* determined $P_m(x)$. It can only be accidential that $P_m(x_k) = f(x_k)$ for some $k > m$, so in general it is not possible to satisfy $n + 1$ interpolation conditions with a polynomial of degree less than n. If $m > n$, we have seen that there is a polynomial $Q(x)$ of degree n such that $Q(x_k) = f(x_k)$ for $k = 0, 1, \ldots, n$. But then if $R(x)$ is *any* polynomial of degree $m - n - 1$, the polynomial

$$R(x)\prod_{k=0}^{n}(x - x_k)$$

is of degree m and so is $P(x)$, the sum of $Q(x)$ and this polynomial. But

$$\begin{aligned} P(x_i) &= Q(x_i) + R(x_i)\prod_{k=0}^{n}(x_i - x_k) \\ &= Q(x_i) \\ &= f(x_i) \quad \text{for} \quad i = 0, 1, \ldots, n. \end{aligned}$$

Thus there are many polynomials which interpolate.

10. The new tables for $x \csc(x)$ and $\csc(x) - \frac{1}{x}$ are

x	$x \csc(x)$	$\csc(x) - \frac{1}{x}$
0.000	1.000000	0.000000
0.005	1.000005	0.001000
0.010	1.000020	0.002000
0.015	1.000038	0.002533
0.020	1.000066	0.003300
0.030	1.000149	0.004967

Interpolating in the table for $x \csc(x)$ at $x = .001$ gives the result 1.000000 with an estimated error 9.5×10^{-7} and requires a polynomial of degree 0. This leads to the answer $\csc(0.001) = 1000.000$. In the second table we obtain 2.000000×10^{-4} with an estimated error of 4.8×10^{-11} using a polynomial of degree 1; this gives $\csc(0.001) = 1000.000$. From standard tables, $\csc(0.001) = 1000.000$ correct to seven significant digits.

11. The plotted solutions of Figures IV.2 and IV.3 show that the higher

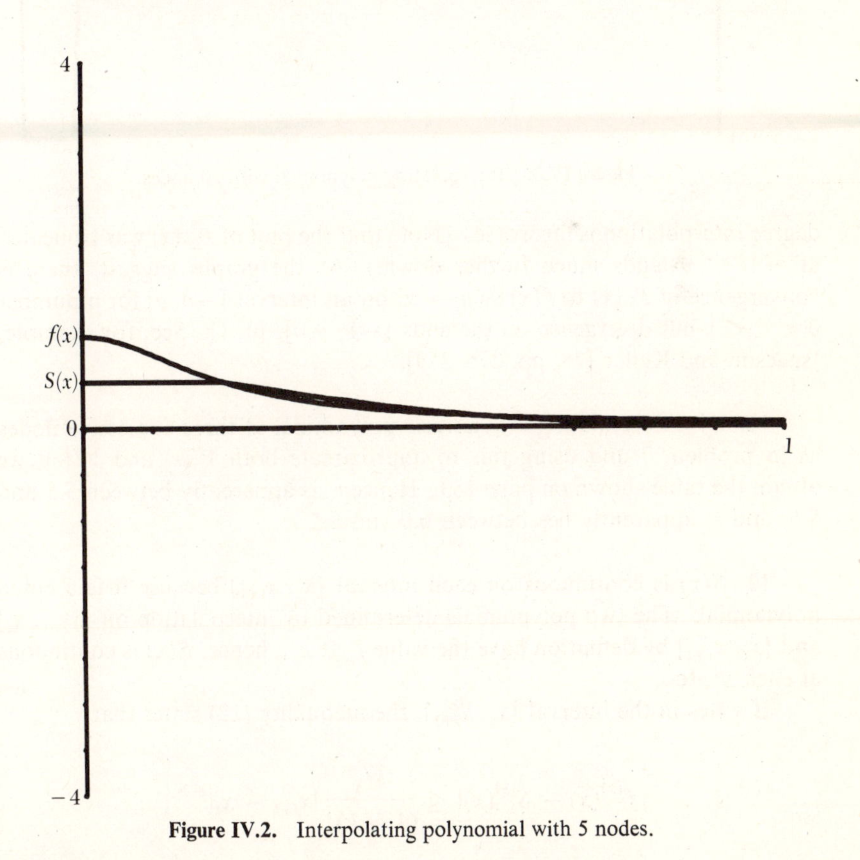

Figure IV.2. Interpolating polynomial with 5 nodes.

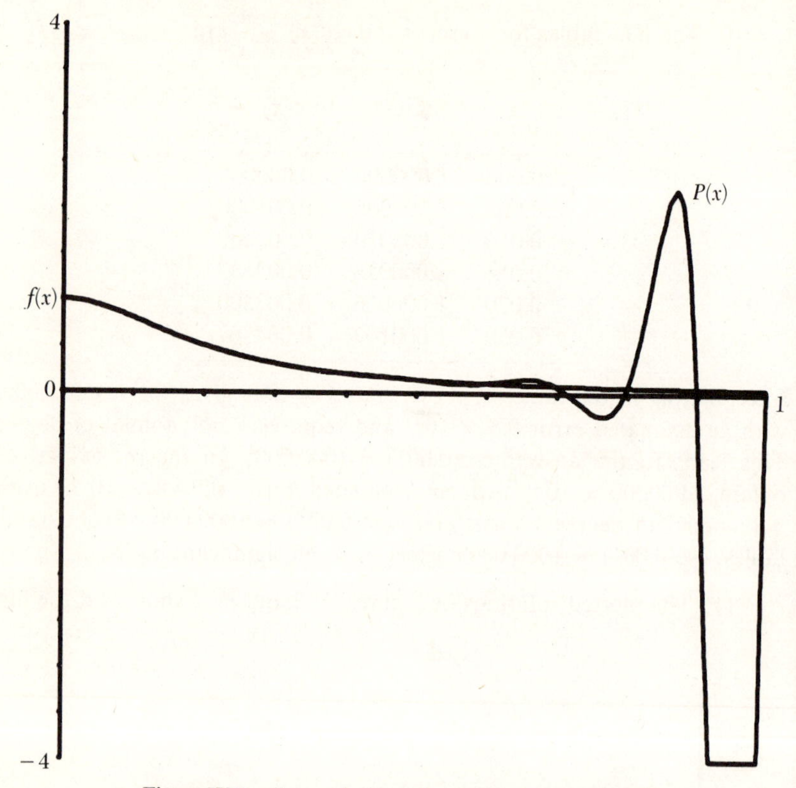

Figure IV.3. Interpolating polynomial with 20 nodes.

degree interpolation is far worse. (Note that the plot of $P_{20}(x)$ was truncated at -4; it extends much further down.) As the graphs suggest, there is convergence of $P_n(x)$ to $f(x)$ as $n \to \infty$ on an interval $[-\delta, \delta]$ for a number $0 < \delta < 1$ but divergence on the ends $[-1, -\delta]$, $[\delta, 1]$. See, for example, Isaacson and Keller [26, pp. 275–279].

12. By differentiating a quadratic polynomial at three successive nodes as in problem 7 and using this to approximate both $V'(r)$ and $V''(r)$, we obtain the table shown on page 183. Hence r_e is apparently between 5.5 and 5.6, and r_i apparently lies between 6.0 and 6.5.

13. $S(x)$ is continuous on each interval $[x_i, x_{i+1}]$ because it is a cubic polynomial. The two polynomials determined by interpolation on $[x_{i-1}, x_i]$ and $[x_i, x_{i+1}]$ by definition have the value f_i at x_i; hence, $S(x)$ is continuous at each x_i, too.

If x lies in the interval $[x_i, x_{i+1}]$, the inequality (12) states that

$$|f^{(k)}(x) - S^{(k)}(x)| \leq \frac{M_4}{(4-k)!} |x_{i+1} - x_i|^{4-k} ;$$

r	$V(r)$	$V'(r)$	$V''(r)$
4.6	32.11		
4.8	9.00	−89.07	264.75
5.0	−3.52	−44.80	178.00
5.1	−7.11	−28.50	148.00
5.2	−9.22	−18.15	59.00
5.3	−10.74	−11.75	69.00
5.4	−11.57	−6.05	45.00
5.5	−11.95	−2.15	33.00
5.6	−12.00	1.10	32.00
5.7	−11.73	3.85	23.00
5.8	−11.23	5.10	2.00
5.9	−10.71	5.50	6.00
6.0	−10.13	5.83	0.53
6.5	−7.15	5.36	−2.40
7.0	−4.77	3.98	−3.12
7.5	−3.17	2.63	−2.28
8.0	−2.14	1.74	−1.27
9.0	−1.03	0.80	−0.62
10.0	−0.54		

but $|x_{i+1} - x_i| \le h$ for each i; hence,

$$|f^{(k)}(x) - S^{(k)}(x)| \le \frac{M_4}{(4-k)!} h^{4-k}.$$

This holds no matter which interval $[x_i, x_{i+1}]$ the point x lies in, so it holds for all x in $[a, b]$.

SOLUTIONS FOR II.1.2

1. We use induction on the degree of the polynomial $P(x)$ to be so represented. Suppose the degree $n = 0$. Now $Q_0(x)$ is of exact degree zero; hence, it is a constant $Q_0(0) \ne 0$. $P(x)$ is a constant too, so

$$P(x) = \alpha_0 Q_0(x)$$

where

$$\alpha_0 = P(0)/Q_0(0).$$

Suppose that all polynomials of degree $n - 1$ can be represented as a linear combination of $Q_0(x), \ldots, Q_{n-1}(x)$. Suppose that $P_n(x)$ is of degree n and

$$P_n(x) = \beta_n x^n + \ldots,$$

$$Q_n(x) = q_n x^n + \cdots$$

where $q_n \neq 0$ since $Q_n(x)$ is of exact degree n. The polynomial

$$R(x) = P_n(x) - \alpha_n Q_n(x)$$

where

$$\alpha_n = \beta_n / q_n$$

is of degree $n - 1$; hence, by assumption,

$$R(x) = \sum_{k=0}^{n-1} \alpha_k Q_k(x).$$

This says that

$$P_n(x) = \sum_{k=0}^{n} \alpha_k Q_k(x)$$

as we desired.

2. If we differentiate the recurrence

$$d_{n+2}(x) = d_{n+1}(x) \equiv 0,$$
$$d_k(x) = c_k + (x - a_{k+1})\, d_{k+1}(x) - b_{k+2} d_{k+2}(x) \qquad k = n, n-1, \ldots, 0,$$
$$R_n(x) = d_0(x),$$

we find

$$d'_{n+2}(x) = d'_{n+1}(x) \equiv 0,$$
$$d'_k(x) = d_{k+1}(x) + (x - a_{k+1})\, d'_{k+1}(x) - b_{k+2}\, d'_{k+2}(x) \quad k = n, n-1, \ldots, 0,$$
$$R'_n(x) = d'_0(x).$$

One then easily shows by induction that for $p > 1$,

$$d^{(p)}_{n+2}(x) = d^{(p)}_{n+1}(x) \equiv 0,$$
$$d^{(p)}_k(x) = p d^{(p-1)}_{k+1}(x) + (x - a_{k+1})\, d^{(p)}_{k+1}(x) - b_{k+2}\, d^{(p)}_{k+2}(x) \;\; k = n, n-1, \ldots 0$$
$$R^{(p)}_n(x) = d^{(p)}_0(x).$$

These recurrences make it quite easy to write a code which will evaluate $R_n(x)$ for a specific x, and along with it the derivatives of orders $1, \ldots, p$. One just carries out all the recurrences simultaneously. With such a code it is easy to convert $R_n(x)$ to power form, since

$$R_n(x) = \sum_{p=0}^{n} \frac{R^{(p)}_n(0)}{p!} x^p$$

from Taylor's expansion. Naturally, one ought to use multiple precision for such a conversion.

3. Using the subroutine LEAST, we find that

$$a_1 = 40.00000,\ a_2 = 40.00000,\ a_3 = 40.00000$$

$$b_2 = 200.0000,\ b_3 = 140.000$$

$$c_0 = 0.475999,\ c_1 = 4.939999 \times 10^{-3}$$

$$c_2 = -4.285722 \times 10^{-5},\ c_3 = 5.833307 \times 10^{-7}.$$

Our least squares polynomial

$$R_3(x) = \sum_{i=0}^{3} c_i Q_i(x)$$

can be put in the power form

$$P_3(x) = \sum_{i=0}^{3} d_i x^i$$

by using the recurrence relation for the orthogonal polynomials to generate the following table.

Coefficient of

	x^0	x^1	x^2	x^3
Q_0	1	0	0	0
Q_1	$-a_1$	1	0	0
Q_2	$-b_2 + a_1a_2$	$-a_1 - a_2$	1	0
Q_3	$a_1b_3 + a_3b_2$ $-a_1a_2a_3$	$-b_3 + a_1a_3$ $+a_2a_3 - b_2 + a_1a_2$	$-a_1$ $-a_2 - a_3$	1

Then

$$d_0 = c_0 - c_1a_1 + c_2(a_1a_2 - b_2) + c_3(a_3b_2 + b_3a_1 - a_1a_2a_3)$$

$$d_1 = c_1 - c_2(a_1 + a_2) + c_3(a_1a_2 + a_1a_3 + a_2a_3 - b_2 - b_3)$$

$$d_2 = c_2 - c_3(a_1 + a_2 + a_3)$$

$$d_3 = c_3$$

Using the above values for the a_i, b_i, and c_i, we find that

$$d_0 = 0.1890000,\ d_1 = 1.097023 \times 10^{-2},$$

$$d_2 = -1.128569 \times 10^{-4},\ d_3 = 5.833307 \times 10^{-7}$$

and so

$$P_3(x) = 0.1890000 + 1.097023 \times 10^{-2}\, x - 1.128569 \times 10^{-4}\, x^2 + 5.833307 \times 10^{-7}\, x^3.$$

The computed values of $P_3(x)$ at the data points, when rounded, agree exactly with the data. A cubic is entirely adequate.

4. The straightforward use of the formulas for orthogonal polynomials can be simplified a little by computing the following quantities:

$$(1, 1) = \sum_{j=1}^{m} w_j \qquad (1, x) = \sum_{j=1}^{m} w_j x_j$$

$$(1, f) = \sum_{j=1}^{m} w_j f_j \qquad (x, x) = \sum_{j=1}^{m} w_j x_j^2$$

$$(x, f) = \sum_{j=1}^{m} w_j x_j f_j$$

By using the rules for our notation it is easy to verify that

$$c_0 = (1, f)/(1, 1)$$

$$a_1 = (1, x)/(1, 1)$$

$$c_1 = \frac{(x, f) - a_1(1, f)}{(x, x) - 2a_1(1, x) + a_1^2(1, 1)}$$

and the linear fit is

$$c_0 - c_1 a_1 + c_1 x.$$

5.

degree of fit	constant term
0	2.844099×10^{-3}
1	2.789656×10^{-3}
2	2.690069×10^{-3}
3	2.698524×10^{-3}
4	2.744146×10^{-3}
5	2.841136×10^{-3}

6. Recall that $w_j > 0$ for all j.

a) $(u, u) = \sum_{j=1}^{m} w_j u(x_j) u(x_j) = \sum_{j=1}^{m} w_j [u(x_j)]^2.$

Each term in the sum is non-negative, and so the sum is non-negative.

b) $\sum_{j=1}^{m} w_j[u(x_j)]^2 = 0$. Clearly the only way the sum can vanish is for each term to vanish; i.e., $u(x_j) = 0, j = 1, 2, \ldots, m$.

c) $(u, v) = \sum_{j=1}^{m} w_j u(x_j)v(x_j) = \sum_{j=1}^{m} w_j v(x_j)u(x_j) = (v, u)$.

d)
$$\left(u, \sum_{i=1}^{p} \alpha_i v_i\right) = \sum_{j=1}^{m}\left[w_j u(x_j) \sum_{i=1}^{p} \alpha_i v_i(x_j)\right]$$
$$= \sum_{i=1}^{p}\left[\alpha_i \sum_{j=1}^{m} w_j u(x_j) v_i(x_j)\right] = \sum_{i=1}^{p} \alpha_i (u, v_i).$$

7. Let EPS_1 denote the RMS error when using TEMP = DBLE(F(I)), and EPS_2 the error when using TEMP = DBLE(F(I)) – R(I). Then

degree of fit	EPS_1	EPS_2
7	3.1×10^{-5}	2.2×10^{-6}
8	4.8×10^{-5}	2.2×10^{-6}
9	7.8×10^{-5}	2.2×10^{-6}
10	1.5×10^{-4}	2.2×10^{-6}

The second form is clearly better computationally.

SOLUTIONS FOR II.1.3

1. The plots for the spline interpolation (Figs. IV.4 and IV.5) exactly correspond to those for the polynomial interpolation problem 11 of II.1.1. The spline fit must perform better for this problem as the number of nodes increases, because Theorem 1 assures us that both the spline and its derivative converge uniformly to $f(x)$ throughout $[-1, 1]$.

2. Using equation (3) we find that

$$\int_{x_0}^{x_n} S(x)\,dx = \sum_{i=0}^{n-1} \int_{x_i}^{x_{i+1}} S(x)\,dx = \sum_{i=0}^{n-1} \int_{x_i}^{x_{i+1}} S_i(x)\,dx$$
$$= \sum_{i=0}^{n-1} \int_{x_i}^{x_{i+1}} \left[\frac{s_i}{6h_i}(x_{i+1} - x)^3 + \frac{s_{i+1}}{6h_i}(x - x_i)^3\right.$$
$$\left. + \left(\frac{f_{i+1}}{h_i} - \frac{s_{i+1}h_i}{6}\right)(x - x_i) + \left(\frac{f_i}{h_i} - \frac{s_i h_i}{6}\right)(x_{i+1} - x)\right] dx$$

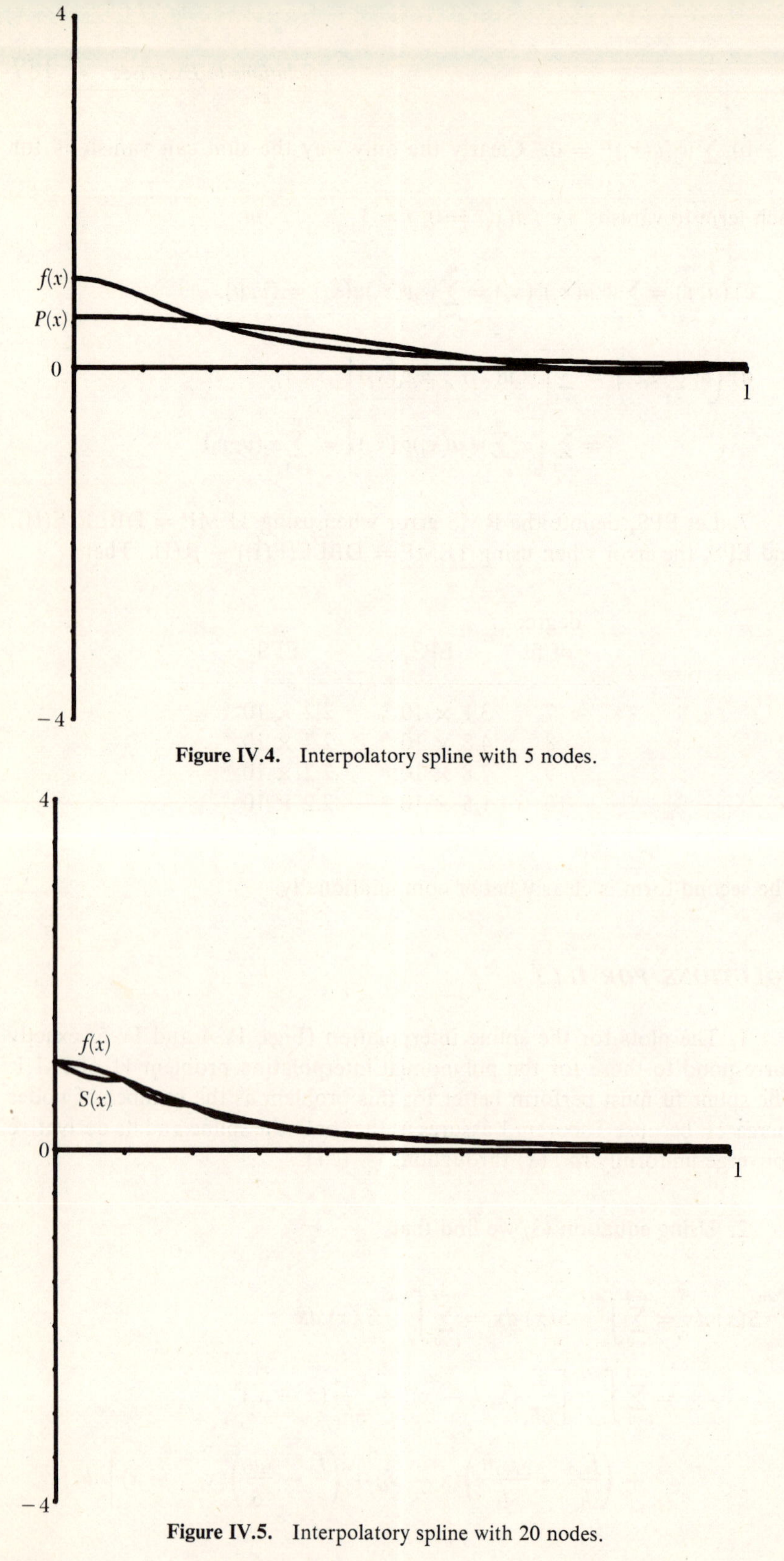

Figure IV.4. Interpolatory spline with 5 nodes.

Figure IV.5. Interpolatory spline with 20 nodes.

$$= \sum_{i=0}^{n-1} \left[-\frac{s_i}{24h_i}(x_{i+1} - x)^4 + \frac{s_{i+1}}{24h_i}(x - x_i)^4 + \frac{1}{2}\left(\frac{f_{i+1}}{h_i} - \frac{s_{i+1}h_i}{6}\right)(x - x_i)^2 - \frac{1}{2}\left(\frac{f_i}{h_i} - \frac{s_i h_i}{6}\right)(x_{i+1} - x)^2 \right]_{x=x_i}^{x=x_{i+1}}$$

$$= \sum_{i=0}^{n-1} \left[\frac{s_{i+1}}{24} h_i^3 + \left(\frac{f_{i+1}}{h_i} - \frac{s_{i+1}}{6} h_i\right)\frac{h_i^2}{2} + \frac{s_i}{24} h_i^3 + \left(\frac{f_i}{h_i} - \frac{s_i h_i}{6}\right)\frac{h_i^2}{2} \right]$$

$$= \sum_{i=0}^{n-1} \left[\frac{h_i}{2}(f_i + f_{i+1}) - \frac{h_i^3}{24}(s_i + s_{i+1}) \right].$$

3. Just implement equation (4) in a straightforward manner after using SPCOEF to calculate the s_i and finding for which i the x of interest satisfies

$$x_i \leq x \leq x_{i+1}.$$

4. The interpolation conditions (ii) make the piecewise polynomial function $S(x)$ continuous. For $S(x)$ to have a continuous derivative it is only necessary to make $S'(x)$ continuous at $x_1, \ldots, x_{n-1}$:

$$S'_{i-1}(x_i) = S'_i(x_i) \qquad i = 2, \ldots, n.$$

But

$$S'_i(x) = b_i + c_i(2x - x_i - x_{i+1})$$

for all i, so this requirement is the equation

$$b_{i-1} + c_{i-1}(2x_i - x_{i-1} - x_i) = b_i + c_i(2x_i - x_i - x_{i+1}).$$

Solving for c_i gives

$$c_i = \frac{b_i - b_{i-1}}{x_{i+1} - x_i} - c_{i-1}\frac{x_i - x_{i-1}}{x_{i+1} - x_i} \qquad i = 2, \ldots, n.$$

We saw in II.1.1 that the divided difference

$$f[x_{k-1}, x_k, x_{k+1}] = \frac{f''(\xi)}{2} \qquad \text{where} \qquad x_{k-1} < \xi < x_{k+1}.$$

Since

$$S''(x) = 2c_k \quad \text{for} \quad x_k \leq x \leq x_{k+1}$$

a natural way to specify all the c_i's is to choose a k with $1 \leq k \leq n - 1$ and use

$$c_k = f[x_{k-1}, x_k, x_{k+1}].$$

Having defined c_k, it is obvious that the recurrence specifies all the remaining c_i's uniquely.

5. It is as easy to obtain all the local maxima and minima as to obtain the global (largest) maximum, so we describe how to do this. First we use SPCOEF to calculate the coefficients s_i describing the cubic interpolatory spline. The extrema occur where $S'(x) = 0$. On each interval $[x_i, x_{i+1}]$ the cubic spline $S(x) = S_i(x)$ is a cubic polynomial and its derivative is the quadratic given by equation (4) of II.1.3,

$$\begin{aligned} S_i'(x) &= -\frac{s_i}{2h_i}(x_{i+1} - x)^2 + \frac{s_{i+1}}{2h_i}(x - x_i)^2 + \frac{f_{i+1} - f_i}{h_i} - \frac{h_i}{6}(s_{i+1} - s_i) \\ &= ax^2 + bx + c \end{aligned}$$

where

$$a = \frac{s_{i+1} - s_i}{2h_i},$$

$$b = \frac{s_i x_{i+1}}{h_i} - \frac{s_{i+1} x_i}{h_i},$$

$$c = \frac{-s_i x_{i+1}^2}{2h_i} + \frac{s_{i+1} x_i^2}{2h_i} + \frac{f_{i+1} - f_i}{h_i} - \frac{h_i}{6}(s_{i+1} - s_i)$$

and, as usual, $h_i = x_{i+1} - x_i$. The two roots of $S_i'(x) = 0$ are

$$r_1 = \frac{-b + \sqrt{b^2 - 4ac}}{2a}, \qquad r_2 = \frac{-b - \sqrt{b^2 - 4ac}}{2a}.$$

First test whether $b^2 - 4ac < 0$. If it is, there are no real roots, so we go on to the next interval. If there are real roots, calculate each and test whether it satisfies

$$x_i \leq r \leq x_{i+1}.$$

If not, it is not a point where $S'(x) = 0$, since $S'(x) = S_i'(x)$ only on $[x_i, x_{i+1}]$ and we forget this root. If it is in the interval, evaluate $S(r)$ using SPLINE and also calculate

$$S''(r) = 2ar + b$$

and print out r, $S(r)$, and $S''(r)$. After testing both roots, go on to the next interval. Repeat this process until all intervals have been treated.

The output gives all the extrema of the spline. Those with $S''(r) > 0$ are local minima, those with $S''(r) < 0$ are local maxima, and those with $S''(r) = 0$ are inflection points. Looking at the values $S(r)$, it is easy to find the global maximum. We found the global maximum to occur at 3986.742 Angstroms.

r	$S(r)$	$S''(r)$
4592.770	0.1416460	5.067142×10^{-6}
3986.742	0.8367515	-8.823904×10^{-6}
3446.295	0.3710922	9.197178×10^{-6}

SOLUTIONS FOR II.2

Unless otherwise stated, all the numerical results given below were obtained by the code SIMP when a relative L_1 accuracy of 10^{-5} was requested. An IFLAG of 1 in every case indicated that the code considered the computation to be successful.

1. One computes the five integrals

$$\int_0^{0.1} e^{t^2}\,dt, \int_{0.1}^{0.2} e^{t^2}\,dt, \ldots, \int_{0.4}^{0.5} e^{t^2}\,dt$$

and their estimated errors $\text{ERROR}_1, \text{ERROR}_2, \ldots, \text{ERROR}_5$. Then, for example,

$$y(0.2) \doteq e^{-(0.2)^2}\left[\int_0^{0.1} e^{t^2}\,dt + \int_{0.1}^{0.2} e^{t^2}\,dt\right] = I_2.$$

We estimate the relative error in $y(0.2)$ to be

$$e^{-(0.2)^2}[\text{ERROR}_1 + \text{ERROR}_2]/I_2.$$

Using the standard tables [1], we calculated by hand the true relative errors given below. Computing $y(0.5)$, say, in this way involves the sum of five errors, whereas doing it directly would involve just one. Because $y(x)$ increases on this range, the relative error remains satisfactory, however.

x	$y(x)$	estimated relative error	true relative error
0.1	9.933585×10^{-2}	3.9×10^{-8}	1.2×10^{-6}
0.2	1.947508×10^{-1}	5.9×10^{-8}	1.2×10^{-6}
0.3	2.826313×10^{-1}	5.1×10^{-8}	8.4×10^{-7}
0.4	3.599431×10^{-1}	4.7×10^{-8}	8.3×10^{-7}
0.5	4.244359×10^{-1}	5.8×10^{-8}	8.4×10^{-7}

2.

p	T	estimated errors (quantity ERROR returned from code)
0.50	3.736762	1.6×10^{-8}
0.75	3.904741	2.8×10^{-8}
1.00	4.187177	7.9×10^{-9}

4. The function $f(x) = [x(x-1)(x-2)(x-3)(x-4)]^2$ has zeros at $x = 0, 1, 2, 3, 4$. In the notation of the text, we see that

$$Q_1(f) = 0, \;\; Q_{11}(f) = 0, \;\; Q_{12}(f) = 0.$$

So the relative L_1 error test

$$|(Q_{11} + Q_{12}) - Q_1| \le \epsilon \, |Q_{11} + Q_{12}|$$

is met after the first level of refinement. The code believes that this positive integral has a value of zero and so "fails." The only way around this particular difficulty is to force the code to do more than one level of refinement before making the decision that the integral has been sufficiently well approximated. Of course, one can generate functions that will make any specific code of this type "fail" in this way.

5. For convenience in notation, we write our polynomial in the form

$$Q[x, x_{i-1}, x_i, x_{i+1}] = a_i x^2 + b_i x + c_i.$$

This can be obtained from the form given in Section II.1.1,

$$Q = f[x_{i-1}] + f[x_{i-1}, x_i](x - x_{i-1}) + f[x_{i-1}, x_i, x_{i+1}](x - x_{i-1})(x - x_i)$$

by letting

$$\begin{aligned} c_i &= f[x_i] - f[x_{i-1}, x_i]x_{i-1} + f[x_{i-1}, x_i, x_{i+1}]x_{i-1}x_i, \\ b_i &= f[x_{i-1}, x_i] - f[x_{i-1}, x_i, x_{i+1}](x_{i-1} + x_i), \\ a_i &= f[x_{i-1}, x_i, x_{i+1}]. \end{aligned}$$

Then for $i = 1, 2, \ldots, n-2$, we use

$$\begin{aligned} \int_{x_i}^{x_{i+1}} f(x)\,dx &\doteq \frac{1}{2}\int_{x_i}^{x_{i+1}} (Q[x, x_{i-1}, x_i, x_{i+1}] + Q[x, x_i, x_{i+1}, x_{i+2}])\,dx \\ &= \frac{a_i + a_{i+1}}{2}\left(\frac{x_{i+1}^3 - x_i^3}{3}\right) + \frac{b_i + b_{i+1}}{2}\left(\frac{x_{i+1}^2 - x_i^2}{2}\right) \\ &\quad + \frac{c_i + c_{i+1}}{2}(x_{i+1} - x_i) \end{aligned}$$

and over the first and last intervals

$$\int_{x_0}^{x_1} f(x)\,dx \doteq \int_{x_0}^{x_1} Q[x, x_0, x_1, x_2]\,dx$$

$$= a_1\left(\frac{x_1^3 - x_0^3}{3}\right) + b_1\left(\frac{x_1^2 - x_0^2}{2}\right) + c_1(x_1 - x_0)$$

and

$$\int_{x_{n-1}}^{x_n} f(x)\,dx \doteq \int_{x_{n-1}}^{x_n} Q[x, x_{n-2}, x_{n-1}, x_n]\,dx$$

$$= a_{n-1}\left(\frac{x_n^3 - x_{n-1}^3}{3}\right) + b_{n-1}\left(\frac{x_n^2 - x_{n-1}^2}{2}\right) + c_{n-1}(x_n - x_{n-1}).$$

6. The change of variables $x = c + t^k$ leads to the integrand

$$g(c + t^k)t^{k-m-1} = g(x)t^{k-m-1}.$$

The integer $k - m - 1$ is non-negative. If it is positive,

$$\frac{d}{dt}[g(x)t^{k-m-1}] = \frac{dg(x)}{dx}\cdot\frac{dx}{dt}\cdot t^{k-m-1} + g(x)\frac{d}{dt}(t^{k-m-1})$$

$$= g'(x)\cdot kt^{k-1}\cdot t^{k-m-1} + g(x)\cdot(k - m - 1)t^{k-m-2}.$$

This obviously exists as a continuous function for $0 \le t \le (b - c)^{1/k}$ when $g'(x)$ exists and is continuous for $c \le x \le b$. Similarly, if $k - m - 1 = 0$, we find

$$\frac{d}{dt}[g(x)] = g'(x)\cdot kt^{k-1}.$$

7. The integrand $\dfrac{1 - e^{-x}}{x}$ apparently has a singularity at $x = 0$. However, since

$$\frac{1 - e^{-x}}{x} = \frac{1 - \left(1 - x + \dfrac{x^2}{2!} - \dfrac{x^3}{3!} + \cdots\right)}{x}$$

$$= 1 - \frac{x}{2!} + \frac{x^2}{3!} - \cdots,$$

we see that

$$\lim_{x\to 0}\frac{1 - e^{-x}}{x} = 1.$$

Thus, there is no mathematical difficulty, although there is a computational one because SIMP will need the value for $x = 0$. In your function subroutine, just evaluate the integrand $h(x)$ by

$$h(x) = \begin{cases} 1 & \text{for} \quad x = 0 \\ \dfrac{1 - e^{-x}}{x} & \text{for} \quad x > 0 \end{cases}$$

and compute with it as usual. For the accuracy request we make, $h(x)$ is not evaluated near enough to $x = 0$ to cause trouble for x values other than 0.

t	$E_1(t)$	estimated error for integral	true answer from tables
1.0	0.2193840	1.8×10^{-7}	0.2193839
2.0	0.0489006	2.6×10^{-7}	0.0489005
3.0	0.0130482	8.2×10^{-7}	0.0130484

8. For convenience in notation, let

$$h(r, \theta, \theta') = \frac{1 - r^2}{1 - 2r\cos(\theta - \theta') + r^2}.$$

Then noting that

$$1 = \frac{1}{2\pi}\int_0^{2\pi} h(r, \theta, \theta')\, d\theta',$$

we have

$$\begin{aligned}
\varphi(r, \theta) &= \frac{1}{2\pi}\int_0^{2\pi} h(r, \theta, \theta') f(\theta')\, d\theta' \\
&\quad + [f(\theta) - f(\theta)] \cdot \frac{1}{2\pi}\int_0^{2\pi} h(r, \theta, \theta')\, d\theta' \\
&= \frac{1}{2\pi}\int_0^{2\pi} h(r, \theta, \theta') f(\theta')\, d\theta' - \frac{1}{2\pi}\int_0^{2\pi} h(r, \theta, \theta') f(\theta)\, d\theta' \\
&\quad + f(\theta)\frac{1}{2\pi}\int_0^{2\pi} h(r, \theta, \theta')\, d\theta' \\
&= \frac{1}{2\pi}\int_0^{2\pi} h(r, \theta, \theta')[f(\theta') - f(\theta)]\, d\theta' + f(\theta) \cdot 1
\end{aligned}$$

which is the desired result. The second form should have better numerical properties for r near 1, since as $\theta' \to \theta$, the numerator $f(\theta') - f(\theta)$ becomes small as the denominator $1 - 2r\cos(\theta - \theta') + r^2$ becomes small. Since the ntegral functions as a small correction to the term $f(\theta)$ and since the integrand is of one sign, the numerator must balance out the effect of the small denominator. Moreover, even if the integral is grossly in error, it has little effect on the result of the sum as $r \to 1$.

To illustrate this, let $\theta = \pi/2$. The solution to our problem is then just $\varphi(r, \theta) = r$. The function φ was evaluated using both forms, the original and the modified, and function counts given as a measure of computational effort.

r	φ original form	φ modified form	function count (original)	function count (modified)
.500000	.500001	.500002	73	61
.750000	.750005	.750000	89	65
.875000	.875004	.875001	137	65
.937500	.937493	.937501	157	61
.968750	.968623	.968750	249	61
.984375	.984287	.984375	1325	69
.992188	.991733	.992187	2033	77

The estimated errors were of the order of 10^{-5} for the first form and 10^{-7} for the second form. The modified form seems more accurate for all r values.

9. Note that $\int_{-\infty}^{\infty} = \int_{-\infty}^{-z} + \int_{-z}^{z} + \int_{z}^{\infty}$. We attempt to bound $\int_{z}^{\infty}$ (which also bounds $\int_{-\infty}^{-z}$) and choose z so that this bound is as small as we desire. For $z > 0.99$,

$$\left|\int_z^\infty \frac{F(\zeta)\,d\zeta}{\cosh\left[\dfrac{(\zeta - x)}{b}\pi\right] + \cos\left(\dfrac{\pi y}{b}\right)}\right| = \left|\int_z^\infty \frac{e^{-100(|\zeta|-0.99)}\,d\zeta}{\cosh\left[\dfrac{(\zeta - x)}{b}\pi\right] + \cos\left(\dfrac{\pi y}{b}\right)}\right|$$

$$\leq e^{99}\int_z^\infty \frac{e^{-100|\zeta|}\,d\zeta}{1 + \cos\left(\dfrac{\pi y}{b}\right)}$$

$$= \frac{e^{99}}{1 + \cos\left(\dfrac{\pi y}{b}\right)}\int_z^\infty e^{-100|\zeta|}\,d\zeta$$

$$= \frac{e^{99}}{100\left(1 + \cos\left(\dfrac{\pi y}{b}\right)\right)}\,e^{-100z}.$$

In our case, when $b = \pi$ and $y = \pi/2$, this bound becomes

$$\tfrac{1}{100}\,e^{-100(z-0.99)}.$$

Using a table of e^{-x}, we see that a z of 1.1 will make this bound approximately equal to 1.7×10^{-7}. So we may replace

$$\int_{-\infty}^{\infty} \frac{F(\zeta)\, d\zeta}{\cosh\left[\frac{(\zeta - x)\pi}{b}\right] + \cos\left(\frac{\pi y}{b}\right)}$$

by

$$\int_{-1.1}^{1.1} \frac{F(\zeta)\, d\zeta}{\cosh\left[\frac{(\zeta - x)\pi}{b}\right] + \cos\left(\frac{\pi y}{b}\right)}$$

with an error not exceeding 3.4×10^{-7}.

To show $\varphi(x, y) = \varphi(-x, y)$, we write

$$\varphi(-x, y) = \frac{1}{b} \sin \frac{\pi y}{b} \int_{-\infty}^{\infty} \frac{F(\zeta)\, d\zeta}{\cosh\left[\frac{(\zeta + x)\pi}{b}\right] + \cos\left(\frac{\pi y}{b}\right)}$$

Make the change of variables $\zeta = -\zeta$ to get

$$\varphi(-x, y) = \frac{1}{b} \sin \frac{\pi y}{b} \int_{+\infty}^{-\infty} \frac{-F(-\zeta)\, d\zeta}{\cosh\left[\frac{(-\zeta + x)\pi}{b}\right] + \cos\left(\frac{\pi y}{b}\right)}.$$

Note that $F(\zeta) = F(-\zeta)$ and $\cosh\left[\frac{(-\zeta + x)\pi}{b}\right] = \cosh\left[\frac{(\zeta - x)\pi}{b}\right]$ and we have

$$\varphi(-x, y) = \varphi(x, y).$$

The resulting plot of $\varphi(x, \pi/2)$ is given in Figure IV.6.

10.

n	q_n	$\int_0^a rf(r)J_0(q_n r)\, dr$	A_n
1	0.9407705	7.556313×10^{-2}	0.1884912
2	3.959371	2.445719×10^{-2}	0.3016393
3	7.686380	6.437087×10^{-4}	1.704085×10^{-4}

The estimated errors returned with the approximate integrals were 7.3×10^{-9}, -1.4×10^{-8}, and 1.0×10^{-8} respectively.

11. The analytical value of the integral $\int_0^a r \sin(\gamma r)\, dr$ is

$$\frac{1}{\gamma^2} [\sin(\gamma a) - (\gamma a)\cos(\gamma a)].$$

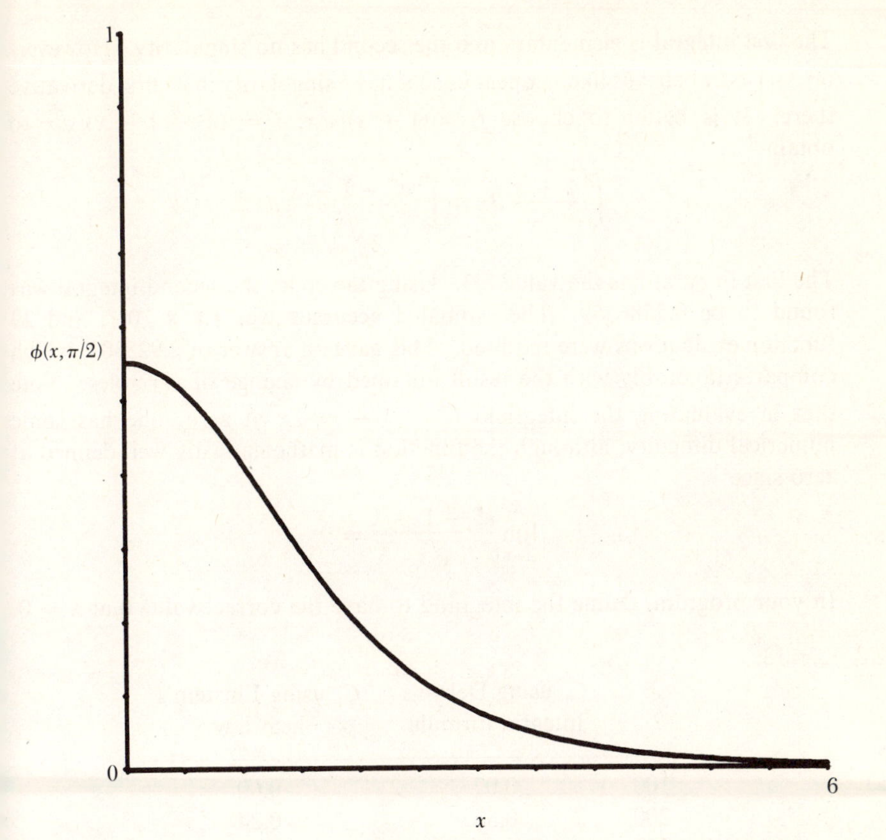

Figure IV.6

n	computed value of integral	estimated error	actual value of integral	A_n
1	0.1579501×10^{-1}	-4.6×10^{-9}	0.1579506×10^{-1}	56.97002
2	$-0.4326664 \times 10^{-3}$	4.7×10^{-9}	$-0.4328331 \times 10^{-3}$	-0.3012316
3	0.1490270×10^{-3}	-7.3×10^{-10}	0.1492598×10^{-3}	0.1008467

k	$t = 10^k$	$\varphi(0.25, t)$
2	10^2	99.44397
3	10^3	94.05487
4	10^4	56.54239
5	10^5	0.3490958

12. Choose $f_1 = \dfrac{1}{\sqrt{x}}, f_2 = \dfrac{e^x - 1}{\sqrt{x}}$ to obtain

$$\int_0^1 \frac{e^x}{\sqrt{x}}\,dx = \int_0^1 \frac{1}{\sqrt{x}}\,dx + \int_0^1 \frac{e^x - 1}{\sqrt{x}}\,dx.$$

The first integral is elementary and the second has no singularity. However $(e^x - 1)/\sqrt{x}$ behaves like $\sqrt{x}$ near 0, so it has a singularity in its first derivative there. It is better to choose $f_1 = (1 + x)/\sqrt{x}$, $f_2 = (e^x - 1 - x)/\sqrt{x}$ to obtain

$$\int_0^1 \frac{1 + x}{\sqrt{x}}\, dx + \int_0^1 \frac{e^x - 1 - x}{\sqrt{x}}\, dx.$$

The first integral has the value 8/3. Using the code, the second integral was found to be 0.2586359. The estimated accuracy was 1.8×10^{-7}, and 29 function evaluations were required. This gave an answer of 2.925302, which compares favorably with the result obtained by change of variables. Note that in evaluating the integrand $(e^x - 1 - x)/\sqrt{x}$ at zero, one has some numerical difficulty, although the function is mathematically well defined at zero since

$$\lim_{x \to 0} \frac{e^x - 1 - x}{\sqrt{x}} = 0.$$

In your program, define the integrand to have the correct value 0 at $x = 0$.

13.

T	C_v using Debye's integral formula	C_v using Einstein's heat law
100	0.07	0.00
200	0.52	0.24
300	1.42	1.18
400	2.38	2.27
500	3.18	3.15
1000	5.01	5.04

14. Any cubic polynomial can be represented exactly by its Taylor series expansion. Let $P(x)$ be a cubic polynomial and $c = \dfrac{b + a}{2}$; then

$$P(x) = P(c) + P'(c)(x - c) + \frac{P''(c)}{2}(x - c)^2 + \frac{P'''(c)}{6}(x - c)^3.$$

Integrating leads to

$$\begin{aligned}\int_a^b P(x)\, dx &= \int_a^b P(c)\, dx + \int_a^b P'(c)(x - c)\, dx \\ &\quad + \int_a^b \frac{P''(c)}{2}(x - c)^2\, dx + \int_a^b \frac{P'''(c)}{6}(x - c)^3\, dx \\ &= \int_a^b P(c)\, dx + \int_a^b \frac{P''(c)}{2}(x - c)^2\, dx.\end{aligned}$$

The integrals involving $(x - c)$ and $(x - c)^3$ vanish because their integrands are odd functions about c. Since

$$b - c = \frac{b-a}{2}, \qquad a - c = \frac{a-b}{2} = -\left(\frac{b-a}{2}\right)$$

some manipulation shows that

$$\int_a^b P(x)\,dx = \frac{b-a}{6}\left[6P(c) + \frac{P''(c)}{4}(b-a)^2\right].$$

If we apply Simpson's rule to $P(x)$ over $a \leq x \leq b$, we have

$$S(P) = \frac{b-a}{6}\,[P(a) + 4P(c) + P(b)].$$

Using the Taylor expansion for $P(x)$, we find

$$\begin{aligned}
S(P) = \frac{b-a}{6}\Bigg[&P(c) + P'(c)(a-c) + \frac{P''(c)}{2}(a-c)^2 \\
&+ \frac{P'''(c)}{6}(a-c)^3 + 4P(c) + P(c) + P'(c)(b-c) \\
&+ \frac{P''(c)}{2}(b-c)^2 + \frac{P'''(c)}{6}(b-c)^3\Bigg] \\
= \frac{b-a}{6}\Bigg[&6P(c) + P'(c)(a-c+b-c) + \frac{P''(c)}{2}((a-c)^2 + (b-c)^2) \\
&+ \frac{P'''(c)}{6}((a-c)^3 + (b-c)^3)\Bigg] \\
= \frac{b-a}{6}\Bigg[&6P(c) + \frac{P''(c)(b-a)^2}{4}\Bigg] \\
= \int_a^b &P(x)\,dx
\end{aligned}$$

which establishes the result.

15. Using the code AVINT as given in [13], the following results were obtained:

quantity measured	integral
glucose	6927.078
insulin	1904.576
glucagon	92688.25
growth hormone	256.6880

SOLUTIONS FOR II.3

1. Scaling $f(x) = 0$ by multiplying by a constant c affects the residual $f(r)$ by the same factor, $cf(r)$. So, a small residual may be due to a scale factor and not the accuracy of the root. Study problem 6, where the scaling makes the coefficients of the polynomial of a nice size but makes the graph very steep near a large root. This has the consequence that a small error in an approximate root leads to a large residual.

2 to 4. All of the problems have simple roots, as can be seen from the fact that the derivatives are non-zero at the roots. The roots are respectively 2.58508424725114, 456.9974921318, 1.30654237418881.

5. Let $f(x) = xe^{-x} - \gamma$. Then $f'(x) = (1 - x)e^{-x}$. For $x < 1$, $f'(x) > 0$ so $f(x)$ is strictly increasing and can have at most one root. Since $f(1) = e^{-1} - \gamma > 0$ and $f(0) = -\gamma < 0$, there is a root in (0, 1). For $x > 1$, $f(x)$ is strictly decreasing. Since $\lim_{x \to \infty} f(x) = -\gamma < 0$, there is a root in $(1, \infty)$. Consequently, there are exactly two roots which are simple and positive. Guessing $x = 0$ and using one step of Newton's method says that the smaller root is approximately γ (which the series shows is a good approximation when γ is small). For $\gamma = 0.06064$, the smaller root is approximately 0.06469262 and the larger is 4.249633.

6.

Root	Bound Problem 7	Bound Problem 8	Residual
3.226×10^{-2}	2.0×10^{-5}	1.8×10^{-4}	-2.5
1.146×10^{3}	8.3	1.1×10^{-3}	-1.1×10^{6}
-1.145×10^{3}	1.4	1.7×10^{-4}	-1.7×10^{5}

In each case the bounds are for relative errors. The residual and bounds were computed in double precision.

7. The expression for a_0 follows from $a_0 = P(0)$ and the factorization of $P(x)$. The expression

$$\left|\frac{P(\sigma)}{a_0}\right|^{1/n} = \left|\prod_{i=1}^{n}\left(\frac{\sigma - r_i}{r_i}\right)\right|^{1/n}$$

is the geometric mean (average value) of the factors $(\sigma - r_i)/r_i$, and the bound just states that the least value of a factor is no more than the average value. If some factor is much larger than the smallest one, the average will be substantially larger than the smallest factor and the bound will be rather poor. If σ is approximately a large root, say r_j, then the factor $(\sigma - r_j)/r_j$ is small. But if some root, say r_k, is much smaller than σ, $|r_k| \ll |\sigma|$, then the factor

$$\left|\frac{\sigma - r_k}{r_k}\right| \doteq \left|\frac{\sigma}{r_k}\right|$$

is quite large and the bound is poor.

8. From the factorization stated in the preceding problem,

$$\ln P(x) = \sum_{j=1}^{n} \ln(x - r_j).$$

Differentiating this gives the desired relation, and the rest of the problem follows the outline in its statement.

If we apply Newton's method with σ as a guess for a root, we get an improved root

$$\sigma^* = \sigma - \frac{P(\sigma)}{P'(\sigma)}.$$

The bound states that

$$\min_j |\sigma - r_j| \leq |\sigma - \sigma^*|$$

so there is a root at least as close to σ as the next iterate.

9. The code is somewhat simpler when based on Newton's method. Newton's method has better behavior at limiting precision than the secant rule, but the use of bisection largely compensates for this. The ultimate rate of convergence of Newton's method is faster than for the secant rule, but each step costs (roughly) twice as much since one must evaluate $f'(x)$ as well as $f(x)$. As a result, the secant rule is a little more efficient. ZEROIN is to be preferred because it does not require the derivative.

10. Basically, one just applies ZEROIN to $u(x)$. It is only necessary to make the proper decision when a pole is present. This is easily dealt with by testing for the magnitudes of the function values and remembering that $f(x)$ has no poles, so one can always use its values to make the proper choice of iterate. This code is more complex than ZEROIN and requires derivatives, but has several important advantages. One can compute roots of even multiplicity and speed up the computation of all multiple roots.

11. Simple roots α have graphs $f(x)$ which cross the axis with a non-zero slope $f'(\alpha) \neq 0$, and multiple roots have graphs tangent to the axis since $f'(\alpha) = 0$. If the slope is small or zero, then $f(x)$ is small in a "large" interval about α and the root is poorly determined.

12. Roots are

$$0.940770563949734,\ 3.95937118501253,\ \text{and}\ 7.08638084796166.$$

13. For a given interval length, the code checks to see if four iterations (four function evaluations) have reduced the interval length by a factor of 1/8; if not, it bisects until it is so reduced. This requires no more than three bisections. So, the absolute maximum of function evaluations required to reduce the interval by a factor of 1/8 is 7. If $|B - C| = 10^{10}$ on input, we ask what is m such that

$$10^{-5} \geq (\tfrac{1}{8})^m \times 10^{10};$$

for in a maximum of $7m$ evaluations the endpoint B is within 10^{-5} of a root. An m of 17 suffices; hence, 119 function evaluations will certainly do the job.

14. An easy way to discuss this problem is to write it as

$$\frac{1}{\omega_0} = \frac{1}{2k} \ln\left(\frac{1 + k}{1 - k}\right) = f(k)$$

for $1 < \dfrac{1}{\omega_0}$. If $\bar{k} > 0$ is a root, then so is $-\bar{k}$ since

$$f(-\bar{k}) = -\frac{1}{2\bar{k}} \ln\left(\frac{1 - \bar{k}}{1 + \bar{k}}\right) = f(\bar{k})$$

from the properties of logarithm. The Taylor series

$$\ln(1 + k) = k - \frac{k^2}{2} + \frac{k^3}{3} - \frac{k^4}{4} + \frac{k^5}{5} - + \cdots,$$

which is valid for $k^2 < 1$, leads to

$$f(k) = 1 + \frac{k^2}{3} + \frac{k^4}{5} + \frac{k^6}{7} + \cdots$$

for $k^2 < 1$. Now $f(0) = 1 < \dfrac{1}{\omega_0}$ and as $k \to 1$, $f(k) \to +\infty$; so for k sufficiently close to 1, $f(k) > \dfrac{1}{\omega_0}$. Clearly, $f(k)$ is strictly increasing in k so there is a unique value $0 < \bar{k} < 1$ such that $f(\bar{k}) = \dfrac{1}{\omega_0}$.

For $\omega_0 = 0.25, 0.50, 0.75$ the roots $\bar{k}$ are approximately 0.9993257, 0.9575040, 0.7755163.

15. If we let $x = 500/2c$, we can rewrite the equation for c as

$$f(x) = 5 \cosh x - 5 - x = 0.$$

Since $\cosh x \geq 1$ for all x, we see that $f(x) > 0$ for all $x < 0$; hence, there are no negative roots. Clearly $f(0) = 0$, but this would correspond to $c = \infty$, which is not admissible in the original problem. Now

$$f'(x) = 5 \sinh x - 1.$$

The function $\sinh x$ is zero for $x = 0$ and strictly increases to $+\infty$ as $x \to \infty$, so $f'(x)$ is negative for $x = 0$ and strictly increases to $+\infty$ as $x \to \infty$. Since $f(0) = 0$, we conclude that $f(x)$ strictly decreases from a value zero at $x = 0$ to a negative minimum value at x_0 where $5 \sinh x_0 - 1 = 0$ and then strictly increases. From

$$\cosh x = \tfrac{1}{2}(e^x + e^{-x})$$

it is easy to see that $f(x) \to +\infty$ as $x \to +\infty$ so $f(x)$ has a unique simple, positive root. It is about 0.3948466, which leads to a root c of about 633.1572 and a value $T = 353.8752$.

16.

x(ft.)	t(sec.)
0.1	0.6565791
0.2	1.236002
0.3	1.918450
0.4	2.694181
0.5	3.556561

17. Assume that x_i, x_{i-1} are near α; then

$$f(x_i) \doteq f(\alpha) + (x_i - \alpha)f'(\alpha) + \frac{(x_i - \alpha)^2}{2} f''(\alpha),$$

$$f(x_{i-1}) \doteq f(\alpha) + (x_{i-1} - \alpha)f'(\alpha) + \frac{(x_{i-1} - \alpha)^2}{2} f''(\alpha).$$

Subtracting α from both sides of (6), using the above expressions, and the fact that $f(\alpha) = 0$, we have

$$(x_{i+1} - \alpha) \doteq (x_i - \alpha) - \frac{\left[f'(\alpha)(x_i - \alpha) + \dfrac{(x_i - \alpha)^2}{2} f''(\alpha)\right](x_i - x_{i-1})}{f'(\alpha)(x_i - x_{i-1}) + \dfrac{f''(\alpha)}{2}[(x_i - \alpha)^2 - (x_{i-1} - \alpha)^2]}$$

or, after some algebraic simplification,

$$x_{i+1} - \alpha \doteq \frac{(x_i - \alpha)(x_{i-1} - \alpha)f''(\alpha)}{f''(\alpha)[(x_i - \alpha) + (x_{i-1} - \alpha)] + 2f'(\alpha)}$$

Writing this in the form

$$\frac{(x_{i+1} - \alpha)}{(x_i - \alpha)(x_{i-1} - \alpha)} \doteq \frac{f''(\alpha)}{f''(\alpha)[(x_i - \alpha) + (x_{i-1} - \alpha)] + 2f'(\alpha)},$$

we see that since $f'(\alpha) \neq 0$, for x_i and x_{i-1} sufficiently close to α

$$\frac{x_{i+1} - \alpha}{(x_i - \alpha)(x_{i-1} - \alpha)} \doteq \frac{f''(\alpha)}{2f'(\alpha)}$$

and the result follows.

18.

Reynold's Number	c_f
10^4	$.7704634 \times 10^{-2}$
10^5	$.4486892 \times 10^{-2}$
10^6	$.2903995 \times 10^{-2}$

19. We use induction to establish the result. It can be easily verified by direct calculation that $\delta_2 = 2$ and $\delta_3 = 3$, which shows that

$$\epsilon_2 \leq \epsilon^{\delta_2} = \epsilon^2,$$
$$\epsilon_3 \leq \epsilon^{\delta_3} = \epsilon^3.$$

Assume that the result holds for $i = k - 1$ and k; then

$$\epsilon_{k+1} \leq \epsilon_k \epsilon_{k-1} \leq \epsilon^{\delta_k} \epsilon^{\delta_{k-1}} = \epsilon^{\delta_k + \delta_{k-1}}.$$

Now

$$\delta_k + \delta_{k-1} = \frac{1}{\sqrt{5}}\left[\left(\frac{1+\sqrt{5}}{2}\right)^{k+1} - \left(\frac{1-\sqrt{5}}{2}\right)^{k+1} + \left(\frac{1+\sqrt{5}}{2}\right)^{k} - \left(\frac{1-\sqrt{5}}{2}\right)^{k}\right]$$
$$= \frac{1}{\sqrt{5}}\left[\left(\frac{1+\sqrt{5}}{2}\right)^{k}\left(\frac{3+\sqrt{5}}{2}\right) - \left(\frac{1-\sqrt{5}}{2}\right)^{k}\left(\frac{3-\sqrt{5}}{2}\right)\right]$$
$$= \frac{1}{\sqrt{5}}\left[\left(\frac{1+\sqrt{5}}{2}\right)^{k}\left(\frac{1+\sqrt{5}}{2}\right)^{2} - \left(\frac{1-\sqrt{5}}{2}\right)^{k}\left(\frac{1-\sqrt{5}}{2}\right)^{2}\right]$$
$$= \delta_{k+1}.$$

So

$$\epsilon_{k+1} \leq \epsilon^{\delta_{k+1}}$$

and the proof is complete.

20. $E/V_0 = .5462489$; $E = 1.190 \times 10^{-11}$ erg.

SOLUTIONS FOR II.4

1. Clearly $y(0) = 0$. Now

$$y'(x) = 0, \qquad 0 \le x \le c$$
$$y'(x) = \tfrac{1}{2}(x - c), \qquad c < x \le b.$$

The limits as $x \to c$ are the same in both cases, so y' is continuous on $0 \le x \le b$. Also, for $0 \le x \le c$,

$$y' = 0 = \sqrt{|0|} = \sqrt{|y|}$$

and for $c \le x \le b$

$$y' = \tfrac{1}{2}(x - c) = \sqrt{|\tfrac{1}{4}(x - c)^2|} = \sqrt{|y|}.$$

Thus, y satisfies the differential equation as an identity.

2. Recall that Dawson's integral is the expression

$$y(x) = e^{-x^2}\int_0^x e^{t^2}\,dt.$$

Now $y(0) = 0$ and differentiation shows that

$$\begin{aligned} y' &= e^{-x^2}e^{x^2} + (-2x)e^{-x^2}\int_0^x e^{t^2}\,dt \\ &= 1 - 2xy. \end{aligned}$$

Obviously, y' is continuous on $[0, b]$ for any finite b.

3. i) $y \equiv 1$; $y(0) = 1$, $y' = 0 = \sqrt{|1 - 1^2|} = \sqrt{|1 - y^2|}$.
   ii) $y = \cosh x$; $y(0) = 1$, $y' = \sinh x$
   $= \sqrt{\cosh^2 x - 1} \equiv \sqrt{|1 - \cosh^2 x|} = \sqrt{|1 - y^2|}$.
   iii) $y = \cos x$; $y(0) = 1$, $y' = -\sin x$. Now the quantity $\sqrt{|1 - \cos^2 x|} \equiv \sqrt{|\sin^2 x|} = |\sin x|$ is non-negative. So the largest interval containing $x = 0$ on which $\cos x$ is a solution is $-\pi \le x \le 0$.

4. Let $g(x)$ and $h(x)$ be continuous. Once we observe that $u(x)$ must be positive because it is an exponential, it is clear that $u(x)$ and $v(x)$ are well defined and differentiable. The derivatives are

$$u' = gu$$
$$v' = h/u$$

which are obviously continuous. Now

$$y(a) = u(a)[A + v(a)]$$
$$= 1 \cdot [A + 0] = A$$

and

$$y' = u'[A + v] + uv'$$
$$= gu[A + v] + u\frac{h}{u}$$
$$= gy + h.$$

So y satisfies the initial value problem.

If we just want values of $y(x)$, a numerical solution is as useful as the "explicit" solution given. If we seek qualitative behavior, the explicit form of the solution would be preferable.

Applying this to the differential equation in problem 2, we have

$$g(x) = -2x, \quad h(x) = 1, \quad A = 0, \quad a = 0$$

so

$$u(x) = e^{\int_0^x -2t\,dt} = e^{-x^2}$$
$$v(x) = \int_0^x \frac{1}{e^{-t^2}}\,dt = \int_0^x e^{t^2}\,dt.$$

Hence,

$$y(x) = e^{-x^2}\int_0^x e^{t^2}\,dt.$$

Using the code RKF with a pure relative error request of 10^{-5}, we obtained the results:

x	$y(x)$ from code	$y(x)$ from tables
0.1	0.9933597×10^{-1}	0.9933599×10^{-1}
0.2	0.1947510	0.1947510
0.3	0.2826315	0.2826317
0.4	0.3599433	0.3599435
0.5	0.4244362	0.4244364

5. a) $f_y = 2y$; the partial derivative is not bounded and so does not satisfy a Lipschitz condition for $0 \leq x \leq \pi/2$.

b) $f_y = -2x$; $|f_y| \leq 2|x| \leq 2b$ for $0 \leq x \leq b$. So f satisfies a Lipschitz condition with constant $L = 2b$.

c) $f_y = 1/x$; $|f_y| \leq 1/1 = 1$ for $1 \leq x \leq 2$; $L = 1$.

d) $f_y = 1/x$; $|f_y|$ is not bounded on $-1 \leq x \leq 1$ and so is not Lipschitzian.

e) $f_y = \cos x \cos y$; $|f_y| \leq 1 \cdot 1 = 1$; $L = 1$.

6. a) Let $Y_1 = u$, $Y_2 = u'$, $Y_3 = u''$, $Y_4 = u'''$; then

$$Y_1' = Y_2,$$
$$Y_2' = Y_3,$$
$$Y_3' = Y_4,$$
$$Y_4' = \cos \alpha t + tY_1 - e^t Y_2.$$

b) Let $Y_1 = u$, $Y_2 = u'$, $Y_3 = v$. Solving the system for u'' and v', we have

$$u'' = t - u - \cos t(e^{\alpha t} - u' - v),$$
$$v' = e^{\alpha t} - u' - v.$$

So

$$Y_1' = Y_2,$$
$$Y_2' = t - Y_1 + Y_2 \cos t + Y_3 \cos t - e^{\alpha t} \cos t,$$
$$Y_3' = -Y_2 - Y_3 + e^{\alpha t}.$$

c) Let $Y_1 = u$, $Y_2 = u'$, $Y_3 = v$. Solving the system for u'' and v', we have

$$u'' = 2t + \frac{3}{4} \cos t - \frac{7u}{4} - v$$

$$v' = 2t - \frac{\cos t}{4} - \frac{3u}{4}.$$

So

$$Y_1' = Y_2,$$

$$Y_2' = 2t + \frac{3}{4} \cos t - \frac{7}{4} Y_1 - Y_3$$

$$Y_3' = 2t - \frac{\cos t}{4} - \frac{3}{4} Y_1.$$

d) Let $Y_1 = x$, $Y_2 = x'$, $Y_3 = y$, $Y_4 = y'$, $Y_5 = z$, $Y_6 = z'$; then

$$Y_1' = Y_2,$$
$$Y_2' = X(t, Y_1, Y_3, Y_5, Y_2, Y_4, Y_6)/m,$$
$$Y_3' = Y_4,$$
$$Y_4' = Y(t, Y_1, Y_3, Y_5, Y_2, Y_4, Y_6)/m,$$
$$Y_5' = Y_6$$
$$Y_6' = Z(t, Y_1, Y_3, Y_5, Y_2, Y_4, Y_6)/m.$$

e) Let $Y_1 = u$, $Y_2 = u'$, $Y_3 = u''$, $Y_4 = u^{(3)}$, $Y_5 = u^{(4)}$, $Y_6 = u^{(5)}$; then

$$\begin{aligned}
Y_1' &= Y_2 \\
Y_2' &= Y_3 \\
Y_3' &= Y_4 \\
Y_4' &= Y_5 \\
Y_5' &= Y_6 \\
Y_6' &= e^t - Y_1 Y_2.
\end{aligned}$$

7. We need to develop a few simple rules about functions which satisfy Lipschitz conditions. If for $a \le x \le b$, the two functions $P(x, y)$ and $Q(x, y)$ satisfy Lipschitz conditions:

$$\begin{aligned}
|P(x, u) - P(x, v)| &\le L_P\,|u - v|, \\
|Q(x, u) - Q(x, v)| &\le L_Q\,|u - v|,
\end{aligned}$$

then their sum $P(x, y) + Q(x, y)$ satisfies a Lipschitz condition, for

$$\begin{aligned}
|[P(x, u) + Q(x, u)] - [P(x, v) + Q(x, v)]| \\
&\le |P(x, u) - P(x, v)| + |Q(x, u) - Q(x, v)| \\
&\le (L_P + L_Q)\,|u - v|.
\end{aligned}$$

Moreover, the composite function $P(x, Q(x, y))$ satisfies a Lipschitz condition since

$$\begin{aligned}
|P(x, Q(x, u)) - P(x, Q(x, v))| &\le L_P|Q(x, u) - Q(x, v)| \\
&\le L_P L_Q |u - v|.
\end{aligned}$$

It is obvious that for any c the function $P(x + c, y)$ satisfies a Lipschitz condition for an appropriate interval in $x + c$. It is also obvious that $cP(x, y)$ satisfies a Lipschitz condition.

Using these rules, it is now easy to look at the Fehlberg formulas (18) and (19) and see that $\Phi(x, y)$ satisfies a Lipschitz condition. To start the argument off, $k_1 = f(x, y)$ is Lipschitzian by assumption. Then so is $\frac{h}{4} k_1$ and the sum $y + \frac{h}{4} k_1$. This implies that the composite function $f\left(x, y + \frac{h}{4} k_1\right)$ and also $k_2 = f\left(x + \frac{h}{4}, y + \frac{h}{4} k_1\right)$ are Lipschitzian. The remainder of the argument continues in this way.

8. To show that $y^{(r)}$ can be computed via the formula

$$y^{(r)} = P_r y + Q_r$$

we use induction. The result is true for $r = 2$ since

$$\begin{aligned} y'' &= P_1'y + P_1y' + Q_1' \\ &= P_1'y + P_1(P_1y + Q_1) + Q_1' \\ &= (P_1' + P_1 \cdot P_1)y + (Q_1' + Q_1P_1) \\ &= P_2y + Q_2. \end{aligned}$$

Assume it to be true for $r = k$; then

$$\begin{aligned} y^{(k+1)} &= P_k'y + P_ky' + Q_k' \\ &= P_k'y + P_k(P_1y + Q_1) + Q_k' \\ &= (P_k' + P_1P_k)y + (Q_k' + Q_1P_k) \\ &= P_{k+1}y + Q_{k+1}. \end{aligned}$$

A fifth order formula is obtained by dropping the remainder term in the exact relation

$$y(x_{n+1}) = y(x_n) + h_n\left(y'(x_n) + y''(x_n)\frac{h_n}{2!} + y^{(3)}(x_n)\frac{h_n^2}{3!} + y^{(4)}(x_n)\frac{h_n^3}{4!} + y^{(5)}(x_n)\frac{h_n^4}{5!}\right) + y^{(6)}(\xi)\frac{h_n^6}{6!}$$

where

$$\begin{aligned} h_n &= x_{n+1} - x_n, \\ y' &= 1 - 2xy, \\ y^{(r)} &= P_ry + Q_r, \qquad r = 2, 3, 4, 5. \end{aligned}$$

The P_r and Q_r are given recursively by

$$\begin{aligned} P_r &= P_{r-1}' + P_1P_{r-1}, \\ Q_r &= Q_{r-1}' + Q_1P_{r-1}, \end{aligned}$$

with

$$P_1 = -2x, \qquad Q_1 = 1.$$

9. Let $y(x) = (x + 1)^2$; then $y(0) = 1$ and $y' = 2(x + 1)$. We see that both initial value problems are satisfied since

$$y' = 2(x + 1)$$

and

$$y' = 2(x+1) = \frac{2(x+1)^2}{x+1} = \frac{2y}{x+1}.$$

Heun's method applied to the first equation yields the recurrence

$$\begin{aligned} y_{k+1} &= y_k + h[\tfrac{1}{2}\cdot 2(x_k+1) + \tfrac{1}{2}\cdot 2(x_k+h+1)] \\ &= y_k + h[2(x_k+1)+h] \end{aligned}$$

and applied to the second, the recurrence

$$\begin{aligned} y_{k+1} &= y_k + h\left[\frac{1}{2}\cdot 2\,\frac{y_k}{x_k+1} + \frac{1}{2}\cdot\frac{2\left(y_k + h\dfrac{2y_k}{x_k+1}\right)}{x_k+h+1}\right] \\ &= y_k + hy_k\left[\frac{2x_k+2+3h}{(x_k+1)(x_k+h+1)}\right]. \end{aligned}$$

Substituting $y_k = (x_k+1)^2$ into the first recurrence gives

$$\begin{aligned} (x_k+h+1)^2 &= (x_k+1)^2 + h[2(x_k+1)+h] \\ &= x_k^2 + 2x_k + 1 + 2x_kh + 2h + h^2 \\ &= (x_k+h+1)^2 \end{aligned}$$

which verifies that the exact solution is generated by this algorithm for this equation. Substitution into the second recurrence yields the equation

$$(x_k+h+1)^2 = (x_k+1)^2 + h(x_k+1)^2\cdot\left[\frac{2x_k+2+3h}{(x_k+1)(x_k+h+1)}\right].$$

Assuming that $x_k+h+1 \neq 0$, this can be put into the form

$$(x_k+h+1)^3 = (x_k+h+1)(x_k+1)^2 + h(x_k+1)(2x_k+2+3h).$$

Some algebraic simplification yields

$$h^3 = 0.$$

This is clearly impossible, so the algorithm cannot be exact when applied to the second equation.

10. Note that the equation can be written in the form

$$\frac{d}{dx}(y'' - y'\sin x - y\cos x) = \ln x.$$

We integrate to obtain

$$y'' - y' \sin x - y \cos x = x \ln x - x + c_2,$$

the first integral relation. Writing this in the form

$$\frac{d}{dx}(y' - y \sin x) = x \ln x - x + c_2,$$

we again integrate to get the second integral relation

$$y' - y \sin x = c_1 + c_2 x + \tfrac{1}{2}x^2 \ln x - \tfrac{3}{4}x^2.$$

To determine c_2, we evaluate the first relation at $x = 1$ and use the initial conditions:

$$A_3 - A_2 \sin(1) - A_1 \cos(1) = -1 + c_2,$$

hence

$$c_2 = A_3 - A_2 \sin(1) - A_1 \cos(1) + 1.$$

Evaluating the second relation at $x = 1$ then gives c_1:

$$A_2 - A_1 \sin(1) = c_1 + c_2 - \tfrac{3}{4},$$

hence

$$c_1 = A_2 - A_1 \sin(1) - c_2 + \tfrac{3}{4}.$$

The results tabulated below were obtained by RKF with a pure relative error request of 10^{-6}. The integral relations were written in the form

$$y''(x) - y'(x) \sin x - y(x) \cos x - c_2 - x \ln x + x = 0, \text{ etc.}$$

and the residuals computed.

x	$y''(x)$	$y'(x)$	$y(x)$	first relation	second relation
1.1	1.109567	1.105459	1.105181	-4.17×10^{-7}	-6.56×10^{-7}
1.2	1.218467	1.221890	1.221457	5.96×10^{-7}	-1.79×10^{-7}
1.3	1.320155	1.348911	1.349911	4.77×10^{-7}	-9.54×10^{-7}
1.4	1.406604	1.485413	1.491553	-5.96×10^{-8}	0.
1.5	1.468271	1.629405	1.647241	-1.19×10^{-7}	9.53×10^{-7}
1.6	1.494283	1.777877	1.817581	-5.36×10^{-7}	0.
1.7	1.472867	1.926679	2.002825	2.38×10^{-7}	9.54×10^{-7}
1.8	1.392103	2.070469	2.202747	2.38×10^{-7}	1.91×10^{-6}
1.9	1.240982	2.202749	2.416533	2.38×10^{-7}	9.54×10^{-7}
2.0	1.010753	2.316022	2.642661	-7.15×10^{-7}	1.91×10^{-6}

Clearly, if the solutions are accurate, the integral relations must be satisfied. If the relations are satisfied, however, the solutions may or may not be accurate. Each solution could be in error and the errors correlated in such a way that the integral relations are satisfied. For example, in the second relation an error of ϵ_1 in $y'(x)$ is cancelled out by an error of $\epsilon_1/\sin x$ in $y(x)$ regardless of its magnitude.

12. Recall that the Euler and Heun methods are given by the formulas

$$y_{k+1} = y_k + hf(x_k, y_k)$$

and

$$\tilde{y}_{k+1} = y_k + \frac{h}{2}[f(x_k, y_k) + f(x_k + h, y_k + hf(x_k, y_k))]$$

respectively. So, an estimate of the local error in Euler's method may be computed from the difference $\tilde{y}_{k+1} - y_{k+1}$; i.e.,

$$\frac{h}{2}[f(x_k + h, y_k + hf(x_k, y_k)) - f(x_k, y_k)].$$

Using the results of problem 4, we find that the solution to the initial value problem

$$y' = 10(y - x), \qquad y(a) = A$$

is given by

$$y = e^{10(x-a)}(A - a - 0.1) + x + 0.1.$$

So the true local error at x_{k+1} is

$$[e^{10(x_{k+1}-x_k)}(y_k - x_k - 0.1) + x_{k+1} + 0.1] - y_{k+1}$$

and the global error is

$$y(x_{k+1}) - y_{k+1} = (x_{k+1} + 0.1) - y_{k+1}.$$

Using a step size of $h = \frac{1}{64}$, we obtained the following table.

x	y_k, Euler's method	estimated local error	actual local error	global error
0.25	0.3500000	0.	0.	0.
0.50	0.6000000	0.	0.	0.
0.75	0.8500000	0.	0.	0.
1.00	1.099994	0.	0.	5.7×10^{-6}
1.25	1.349905	-1.0×10^{-6}	-1.9×10^{-6}	9.4×10^{-5}
1.50	1.598994	-1.1×10^{-5}	-1.1×10^{-5}	1.0×10^{-3}
1.75	1.839715	-1.1×10^{-4}	-1.1×10^{-4}	1.0×10^{-2}
2.00	1.995004	-1.1×10^{-3}	-1.2×10^{-3}	1.0×10^{-1}
2.25	1.278448	-1.1×10^{-2}	-1.2×10^{-2}	1.1
2.50	-8.335561	-1.2×10^{-1}	-1.2×10^{-1}	1.0×10^{1}
2.75	-108.7510	-1.2	-1.2	1.1×10^{2}
3.00	-1135.824	-1.4×10^{1}	-1.5×10^{1}	9.9×10^{2}

This initial value problem furnishes us with a good example of the fact that a small local error at every step does not imply small global errors. For example, in the actual computations between 1.75 and 2.00 there are 16 integration steps. The sum of the 16 local errors (not shown in the above table) is approximately -9.8×10^{-3}. The global error changed from 1.0×10^{-2} to 1.0×10^{-1}.

14. The equation $x'' + (x^2 - 1)x' + x = 0$ is of the form treated by Liénard with $f(x) = x^2 - 1$. The indefinite integral $G(x) = x^3/3 - x$, so the Liénard variables are

$$Y_1(t) = x(t),$$
$$Y_2(t) = x'(t) + G(x(t)).$$

The differential equation is written as

$$Y_1'(t) = Y_2(t) + Y_1(t) - Y_1^3(t)/3,$$
$$Y_2'(t) = -Y_1(t).$$

To plot the solution in the phase plane we must plot $x'(t) = Y_2(t) - G(x(t)) = Y_2(t) - G(Y_1(t))$ against $x(t) = Y_1(t)$. The plot shown in Figure IV.7 is for $Y_1(0) = -1$, $Y_2(0) = 1$ and $0 \le t \le 15$. The closed curve was traced over in the computation, so (to plotting accuracy) the limit cycle was obtained.

We remark that there is no advantage to be gained in going to the Liénard variables here; we have done so just to illustrate the transformation.

15. Clearly, $f(x, y) = 2\,|x|\,y$ is continuous on $[-1, 1]$. We have $f_y = 2\,|x|$ and $|f_y| \le 2$, and so f satisfies a Lipschitz condition with constant $L = 2$. The hypotheses of Theorem 1 are satisfied. Let $y(x)$ be defined by

$$y = \begin{cases} e^{x^2}, & x \ge 0 \\ e^{-x^2}, & x < 0; \end{cases}$$

then $y(-1) = e^{-1}$ and

$$y' = \begin{cases} 2xe^{x^2}, & x \ge 0 \\ -2xe^{-x^2}, & x < 0. \end{cases}$$

Thus, y' is continuous on $[-1, 1]$. (The continuity at $x = 0$ follows by letting $x \to 0$ from both the right and the left; in each case $y'(x) \to 0$.) Also

$$y' = \begin{cases} 2xe^{x^2} = 2\,|x|\,y, & x \ge 0 \\ -2xe^{-x^2} = 2\,|x|\,y, & x < 0 \end{cases}$$

and so y satisfies the differential equation.

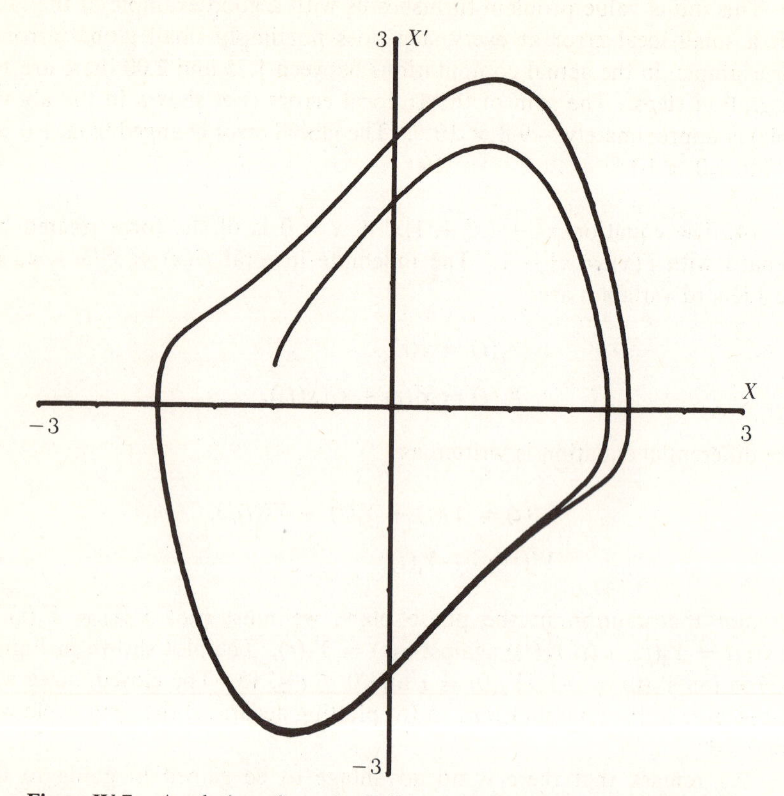

Figure IV.7. A solution of van der Pol's equation ($\epsilon = 1$) in the phase plane.

We see that y does not have two continuous derivatives on $[-1, 1]$ since

$$y'' = \begin{cases} 2e^{x^2} + 4x^2e^{x^2}, & x \geq 0 \\ -2e^{-x^2} + 4x^2e^{-x^2}, & x < 0 \end{cases}$$

is not continuous at $x = 0$. (As $x \to 0$ from the right, $y''(x) \to 2$; as $x \to 0$ from the left, $y''(x) \to -2$.) Euler's method is convergent for this problem (this follows from the fact that y'' is bounded on $[-1, 1]$) but not $0(h)$. Higher order Runge-Kutta methods will not improve convergence.

By splitting the problem at $x = 0$, we see that $y(x)$ is infinitely differentiable on each subinterval $[-1, 0]$ and $[0, 1]$, and so Euler's method is of order h for each subproblem. If, in solving the original problem, one places a mesh point at $x = 0$, this is clearly equivalent to solving the two problems separately.

In using the steps $h = \frac{2}{2^k}$, $x = 0$ is always a mesh point and one gets convergence of order h. This is not the case, however, when $h = \frac{2}{3^k}$.

16. This problem was solved with RKF and a pure absolute error tolerance of 10^{-5} to get the results below. The relations were written in the form

$$sn^2(x) + cn^2(x) - 1 = 0, \text{ etc.}$$

and the residuals computed.

x	$y_1(x)$	$y_2(x)$	$y_3(x)$
1.862640	0.9999995	$-0.3576279 \times 10^{-6}$	0.7000002
3.725281	$-.4053116 \times 10^{-5}$	-1.000000	0.9999996
5.587922	-0.9999996	0.4529953×10^{-5}	0.6999995

x	residual in relation 1	residual in relation 2	residual in relation 3
1.862640	$-.95 \times 10^{-6}$	$-.24 \times 10^{-6}$	$-.30 \times 10^{-6}$
3.725281	0.	$-.83 \times 10^{-6}$	$-.83 \times 10^{-6}$
5.587922	$-.83 \times 10^{-6}$	$-.12 \times 10^{-5}$	$-.72 \times 10^{-6}$

17. Differentiating the first equation, we obtain

$$Y_1'' = -\frac{x}{2} Y_1' - \tfrac{1}{2} Y_1 + Y_2'.$$

Substitute the second equation for Y_2' to get

$$Y_1'' = -\frac{x}{2} Y_1' - \tfrac{1}{2} Y_1 + (\tfrac{1}{2} - \tfrac{3}{4} x^2) Y_1 - \frac{x}{2} Y_2.$$

Solve the first equation for Y_2 and substitute it to arrive at

$$Y_1'' = -\frac{x}{2} Y_1' - \tfrac{1}{2} Y_1 + (\tfrac{1}{2} - \tfrac{3}{4} x^2) Y_1 - \frac{x}{2}\left(Y_1' + \frac{x}{2} Y_1\right)$$

$$= -xY_1' - x^2 Y_1.$$

18. The code RKF was used with pure relative error tolerance of 10^{-5} to compute the results given.

x	$y(x)$	$\tan x$
0.1	0.1003345	0.1003346
0.2	0.2027097	0.2027100
0.3	0.3093357	0.3093363
0.4	0.4227924	0.4227933
0.5	0.5463015	0.5463026
0.6	0.6841356	0.6841370
0.7	0.8422869	0.8422886
0.8	1.029636	1.029638
0.9	1.260155	1.260159
1.0	1.557403	1.557407
1.1	1.964751	1.964756
1.2	2.572136	2.572143
1.3	3.602075	3.602078
1.4	5.797865	5.797804
1.5	14.10192	14.10085
1.556	67.54091	67.35684

A value IFLAG = 4 was returned at the point $x = 1.556$, indicating that the code needed to take too small a step for the computer word length (here an IBM 360/67, which has a relatively short word).

19. When the code tries to hit an output point, the step size may be as large as $1.25 \times h_{\text{max}}$.

20. If the code did not stretch its step to hit the output point B, it could land short of but arbitrarily close to B. On the next step this could cause a return of IFLAG = 4 because the step is too small or at least an inefficiently small step.

If at a given point x, we let h denote the step the code estimates will be successful, the code actually uses $H = 0.8\,h$ to make a failure unlikely. When stretching H to reach an output point, we do not allow it to become larger than $h = 1.25\,H$, which is reasonable since the code does think it likely that such a step will be successful.

SOLUTIONS FOR III

1. In matrix form the system is

$$\begin{pmatrix} -3 & 8 & 5 \\ 2 & -7 & 4 \\ 1 & 9 & -6 \end{pmatrix} \begin{pmatrix} x_1 \\ x_2 \\ x_3 \end{pmatrix} = \begin{pmatrix} 6 \\ 9 \\ 1 \end{pmatrix}.$$

To scale, we multiply the first equation by 1/8, the second by 1/7, and the third by 1/9 to obtain the equivalent system

$$-\tfrac{3}{8}x_1 + x_2 + \tfrac{5}{8}x_3 = \tfrac{6}{8}, \tag{1a}$$

$$\tfrac{2}{7}x_1 - x_2 + \tfrac{4}{7}x_3 = \tfrac{9}{7}, \tag{1b}$$

$$\tfrac{1}{9}x_1 + x_2 - \tfrac{6}{9}x_3 = \tfrac{1}{9}. \tag{1c}$$

The partial pivoting strategy requires no row interchange in the first step of the elimination process, and so we eliminate x_1 from the second and third equations to obtain

$$-\tfrac{5}{21}x_2 + \tfrac{22}{21}x_3 = \tfrac{39}{21},$$

$$\tfrac{35}{27}x_2 - \tfrac{13}{27}x_3 = \tfrac{3}{9}.$$

As $\left|\frac{35}{27}\right| > \left|\frac{-5}{21}\right|$, we interchange to get

$$\tfrac{35}{27}x_2 - \tfrac{13}{27}x_3 = \tfrac{3}{9}, \tag{2a}$$

$$-\tfrac{5}{21}x_2 + \tfrac{22}{21}x_3 = \tfrac{39}{21} \tag{2b}$$

and eliminate x_2 from (2b) to get

$$\tfrac{141}{7}x_3 = \tfrac{282}{7}. \tag{3}$$

By back substitution, we find using (3) that $x_3 = 2$, and then using (2a),

$$\tfrac{35}{27}x_2 - \tfrac{13}{27} \times 2 = \tfrac{3}{9},$$

that $x_2 = 1$ and using (1a),

$$-\tfrac{3}{8}x_1 + 1 \times 1 + \tfrac{5}{8} \times 2 = \tfrac{6}{8},$$

that $x_1 = 4$. So the solution vector is

$$\mathbf{x} = \begin{pmatrix} 4 \\ 1 \\ 2 \end{pmatrix}.$$

As a check, we form the residuals

$$r_1 = 6 - (-3 \times 4 + 8 \times 1 + 5 \times 2) = 0,$$

$$r_2 = 9 - (2 \times 4 - 7 \times 1 + 4 \times 2) = 0,$$

$$r_3 = 1 - (1 \times 4 + 9 \times 1 - 6 \times 2) = 0.$$

2.

$$\begin{pmatrix} 0 & 1 & 1 \\ 1 & -5 & 3 \\ 2 & 1 & -4 \end{pmatrix} \begin{pmatrix} x_1 \\ x_2 \\ x_3 \end{pmatrix} = \begin{pmatrix} 0 \\ 0 \\ -1 \end{pmatrix}, \qquad \mathbf{x} = \begin{pmatrix} -\frac{8}{21} \\ -\frac{1}{21} \\ \frac{1}{21} \end{pmatrix}, \qquad \mathbf{r} = \begin{pmatrix} 0 \\ 0 \\ 0 \end{pmatrix}.$$

3. (a) $\mathbf{x} = \begin{pmatrix} 4.07 \\ 1.02 \\ 2.00 \end{pmatrix}, \qquad \mathbf{r} = \begin{pmatrix} .050 \\ .000 \\ -.250 \end{pmatrix};$

(b) $\mathbf{x} = \begin{pmatrix} 3.97 \\ .991 \\ 1.99 \end{pmatrix}, \qquad \mathbf{r} = \begin{pmatrix} .032 \\ .037 \\ .051 \end{pmatrix};$

(c) $\mathbf{x} = \begin{pmatrix} 4.00 \\ 1.00 \\ 2.00 \end{pmatrix}, \qquad \mathbf{r} = \begin{pmatrix} .000 \\ .000 \\ .000 \end{pmatrix}.$

To illustrate the computations, we carry out the solution of (c). Scaling the original system, we have

$$\begin{aligned} -.375x_1 + 1.00x_2 + .625x_3 &= .750, \\ .286x_1 - 1.00x_2 + .571x_3 &= 1.29, \\ .111x_1 + 1.00x_2 - .667x_3 &= .111. \end{aligned} \tag{1}$$

No interchanges are required, as $|-.375|$ is larger than $|.286|$ and $|.111|$, and so eliminating x_1 gives the two new equations

$$\begin{aligned} -.237x_2 + 1.05x_3 &= 1.86, \\ 1.30x_2 - .482x_3 &= .333. \end{aligned} \tag{2}$$

Interchanging equations and eliminating x_2 gives

$$.962x_3 = 1.92. \tag{3}$$

Back substitution using equations (3), (2) and (1) in that order yields

$$x_3 = 2.00,\ x_2 = 1.00,\ x_1 = 4.00$$

As discussed in the text, the residual calculation is done in double precision. The arithmetic then is all exact, and so we need not be concerned with the order of the arithmetic operations. Hence

$$
\begin{aligned}
r_1 &= 6.00 - (-3.00 \times 4.00 + 8.00 \times 1.00 + 5.00 \times 2.00) \\
&= 6.00 - (-12.0000 + 8.00000 + 10.0000) \\
&= 6.00 - (6.00000) \\
&= 6.00000 - 6.00000 \\
&= 0.00000
\end{aligned}
$$

and similarly

$$r_2 = r_3 = 0.00000.$$

4. Systems (a), (b), and (c) are singular. For (d) we find that

$$\mathbf{x} = \begin{pmatrix} -5 \\ 6 \\ -2 \\ 7 \end{pmatrix}.$$

5. For problem 1, $\mathbf{x} = \begin{pmatrix} 3.999997 \\ 0.9999997 \\ 0.1999999 \end{pmatrix}$ with maximum residual $R = 7.4 \times 10^{-6}$. For problem 2, $\mathbf{x} = \begin{pmatrix} -0.3809524 \\ -0.04761906 \\ 0.04761903 \end{pmatrix}$ with $R = 1.2 \times 10^{-7}$.

6. For $V = 50, 100, 150$, the solution vectors v are respectively

$$\mathbf{v} = \begin{pmatrix} 34.99995 \\ 25.99995 \\ 19.99995 \\ 15.49997 \\ 10.99998 \\ 4.999990 \end{pmatrix}, \begin{pmatrix} 69.99994 \\ 51.99991 \\ 39.99989 \\ 30.99992 \\ 21.99995 \\ 9.999982 \end{pmatrix}, \begin{pmatrix} 104.9999 \\ 77.99985 \\ 59.99980 \\ 46.49986 \\ 32.99991 \\ 14.99997 \end{pmatrix}.$$

with maximum residuals $R = 2.6 \times 10^{-4}$, 6.6×10^{-4}, 9.0×10^{-4}.

7.

$$
\begin{aligned}
&R_E = 223.0000, && R_A = 177.0000, && C_h = 56.76028, \\
&D_h = -56.76028, && C_v = 29.33987, && D_v = -252.3399, \\
&B_v = -147.6603, && B_h = 56.76028
\end{aligned}
$$

with maximum residual $R = 1.2 \times 10^{-3}$.

Solve equations 1, 2, 6 and 7 for the quantities R_E, R_A, D_v and B_v respectively to obtain

$$R_E = 223.0000$$
$$R_A = 177.0000$$
$$D_v = -252.3397$$
$$B_v = -147.6603.$$

Then solve equation 4 for C_v:

$$C_v = 29.33970$$

From equation 5

$$C_h = 56.76069$$

and from 3

$$D_h = -56.76069$$

and from 8

$$B_h = 56.76069.$$

8. $R_1 = 51.67000$, $R_2 = 26.66000$, $R_3 = 31.67000$ with maximum residual $R = 0$.

9. Just to illustrate the idea, we consider the case $n = 3$; then $h = \frac{1}{3}$ and

$$x_1 = \tfrac{1}{3},\ x_2 = \tfrac{2}{3},\ x_3 = 1.$$

Since we have $g(x_i) = 1$, $K(x_i, x_j) = x_i + x_j$, $\lambda = 2$, the equations

$$f(x_i)[1 - \lambda h K(x_i, x_i)] - \lambda h \sum_{\substack{j \neq i \\ j=1}}^{N} K(x_i, x_j) f(x_j) = g(x_i), \qquad i = 1, 2, 3$$

become

$$[1 - 2 \times \tfrac{1}{3} \times (\tfrac{1}{3} + \tfrac{1}{3})] \times f(\tfrac{1}{3}) - [2 \times \tfrac{1}{3} \times (\tfrac{1}{3} + \tfrac{2}{3})] \times f(\tfrac{2}{3})$$
$$- [2 \times \tfrac{1}{3} \times (\tfrac{1}{3} + 1)] \times f(1) = 1,$$
$$[-2 \times \tfrac{1}{3} \times (\tfrac{2}{3} + \tfrac{1}{3})] \times f(\tfrac{1}{3}) + [1 - 2 \times \tfrac{1}{3} \times (\tfrac{2}{3} + \tfrac{2}{3})] \times f(\tfrac{2}{3})$$
$$- [2 \times \tfrac{1}{3} \times (\tfrac{2}{3} + 1)] \times f(1) = 1,$$
$$[-2 \times \tfrac{1}{3} \times (1 + \tfrac{1}{3})] \times f(\tfrac{1}{3}) - [2 \times \tfrac{1}{3} \times (1 + \tfrac{2}{3})] \times f(\tfrac{2}{3})$$
$$+ [1 - 2 \times \tfrac{1}{3} \times (1 + 1)] \times f(1) = 1.$$

This is a linear system in the variables $f(\frac{1}{3}), f(\frac{2}{3}), f(1)$. Using matrix notation it can be written in the form

$$\begin{pmatrix} 1 - 2 \times \frac{1}{3} \times (\frac{1}{3} + \frac{1}{3}) & -2 \times \frac{1}{3} \times (\frac{1}{3} + \frac{2}{3}) & -2 \times \frac{1}{3} \times (\frac{1}{3} + 1) \\ -2 \times \frac{1}{3} \times (\frac{2}{3} + \frac{1}{3}) & 1 - 2 \times \frac{1}{3} \times (\frac{2}{3} + \frac{2}{3}) & -2 \times \frac{1}{3} \times (\frac{2}{3} + 1) \\ -2 \times \frac{1}{3} \times (1 + \frac{1}{3}) & -2 \times \frac{1}{3} \times (1 + \frac{2}{3}) & 1 - 2 \times \frac{1}{3} \times (1 + 1) \end{pmatrix}$$

$$\times \begin{pmatrix} f(\frac{1}{3}) \\ f(\frac{2}{3}) \\ f(1) \end{pmatrix} = \begin{pmatrix} 1 \\ 1 \\ 1 \end{pmatrix}.$$

For the case $n = 40$, the approximate solutions are accurate to about one significant digit. This is obviously not an efficient way to solve integral equations. See [21].

10. (a) The code LASSOL was used to find the coefficients a_i, $i = 0, 1, \ldots, 5$:

$$a_0 = 1.000000,\ a_1 = 1.000072,\quad a_2 = .4991317,$$

$$a_3 = .1702622,\ a_4 = .03501472,\ a_5 = .01380143,$$

with a maximum residual $R = 2.2 \times 10^{-6}$. The polynomial values compared quite well with the corresponding values of e^x, as the following table indicates.

x	$P_5(x)$	e^x
.12	1.127496	1.127497
.23	1.258596	1.258599
.45	1.568309	1.568312
.67	1.954234	1.954237
.89	2.435130	2.435129

(b) $A = 1.496881$, $B = 4.161899 \times 10^5$, $C = 2.323399 \times 10^{11}$ were found using LASSOL with maximum residual $R = 8.9 \times 10^{-5}$. In matrix form, our problem is

$$\begin{pmatrix} 1.000000 & .6249996 \times 10^{-7} & .3906245 \times 10^{-14} \\ 1.000000 & .4938275 \times 10^{-7} & .2438654 \times 10^{-14} \\ 1.000000 & .4000002 \times 10^{-7} & .1600000 \times 10^{-14} \end{pmatrix} \begin{pmatrix} A \\ B \\ C \end{pmatrix} = \begin{pmatrix} 1.5238 \\ 1.5180 \\ 1.5139 \end{pmatrix}.$$

Again using LASSOL, we find that the coefficient matrix has the inverse

$$\begin{pmatrix} 6.692861 & -20.31274 & 14.61989 \\ -3.028524 \times 10^8 & 8.328233 \times 10^8 & -5.299715 \times 10^8 \\ 3.388265 \times 10^{15} & -8.125116 \times 10^{15} & 4.736851 \times 10^{15} \end{pmatrix}$$

The elements of the inverse matrix are quite large, indicating that the problem is ill conditioned.

11. For the first **p**, the displacements are

$$\mathbf{x} = \begin{pmatrix} 5.499954 \times 10^{-6} \\ -1.653208 \times 10^{-4} \\ -3.717327 \times 10^{-6} \\ -4.737385 \times 10^{-5} \\ 3.713940 \times 10^{-5} \\ -1.211510 \times 10^{-4} \\ 6.434067 \times 10^{-5} \\ 6.361977 \times 10^{-5} \end{pmatrix}$$

and the corresponding forces

$$\mathbf{f} = \begin{pmatrix} -6.189920 \times 10^{-1} \\ -9.217280 \times 10^{-2} \\ 1.201534 \times 10^{-2} \\ 3.713940 \times 10^{-1} \\ 2.720127 \times 10^{-1} \\ -7.209003 \times 10^{-3} \\ -1.325095 \times 10^{-1} \\ -4.653707 \times 10^{-1} \\ 1.656370 \times 10^{-1} \\ -1.421215 \times 10^{-1} \end{pmatrix}$$

12. Using LASSOL the solution is

$$\mathbf{x} = \begin{pmatrix} -6.861310 \\ -5.182483 \times 10^{-1} \\ -2.087591 \end{pmatrix}, \qquad R = 1.2 \times 10^{-5}.$$

Again using the code we find that

$$A^{-1} = \begin{pmatrix} -4.379574 \times 10^{-2} & -6.788319 \times 10^{-1} & -4.890509 \times 10^{-1} \\ 5.839416 \times 10^{-2} & -9.489048 \times 10^{-2} & -1.459854 \times 10^{-2} \\ 8.029199 \times 10^{-2} & -2.554744 \times 10^{-1} & -2.700729 \times 10^{-1} \end{pmatrix}$$

$$A^{-1}\mathbf{b} = \begin{pmatrix} -6.861311 \\ -5.182478 \times 10^{-1} \\ -2.087590 \end{pmatrix}$$

which has the maximum residual $R = 4.1 \times 10^{-5}$.

This second method of solution not only produces larger residuals (4.1×10^{-5} as compared to 1.2×10^{-5}) but requires more work. Counting only multiplications and divisions as operations, we see that to solve the given problem by elimination requires 17 operations, while to solve by first computing A^{-1} and then forming the product $A^{-1}\mathbf{b}$ requires 44 operations (if one calculates A^{-1} as we do here; one can economize somewhat).

13. Suppose that the matrix A has an inverse denoted by A^{-1} (this means that the matrix A is non-singular and the problem $A\mathbf{x} = \mathbf{b}$ has a unique solution $\mathbf{x}$ for all right hand sides $\mathbf{b}$). Denote the elements of A^{-1} by α_{ij}. Consider the problem

$$A\mathbf{x}^1 = \mathbf{b}^1$$

which we know from the theory has the solution

$$\mathbf{x}^1 = A^{-1}\mathbf{b}^1 = \begin{pmatrix} \alpha_{11} & \alpha_{12} & \cdots & \alpha_{1n} \\ \alpha_{21} & \alpha_{22} & \cdots & \alpha_{2n} \\ \cdot & & & \\ \cdot & & & \\ \cdot & & & \\ \alpha_{n1} & \alpha_{n2} & \cdots & \alpha_{nn} \end{pmatrix} \begin{pmatrix} 1 \\ 0 \\ \cdot \\ \cdot \\ \cdot \\ 0 \end{pmatrix}$$

$$= \begin{pmatrix} \alpha_{11} \\ \alpha_{21} \\ \alpha_{31} \\ \cdot \\ \cdot \\ \cdot \\ \alpha_{n1} \end{pmatrix}.$$

So the vector $\mathbf{x}^1$ is the first column of A^{-1}. In general, if we solve

$$A\mathbf{x}^i = \mathbf{b}^i,$$

we find that

$$\mathbf{x}^i = A^{-1}\mathbf{b}^i = \begin{pmatrix} \alpha_{1i} \\ \alpha_{2i} \\ \cdot \\ \cdot \\ \cdot \\ \alpha_{ni} \end{pmatrix}$$

which is the ith column of A^{-1}. Thus, if we form the matrix X with the n columns $\mathbf{x}^1, \mathbf{x}^2, \ldots, \mathbf{x}^n$, we see that $X = A^{-1}$.

14. Using the result of problem 13, we solve the two problems

$$\begin{pmatrix} 1, & -1 \\ 1, & -1+10^{-5} \end{pmatrix} \begin{pmatrix} x_1^1 \\ x_2^1 \end{pmatrix} = \begin{pmatrix} 1 \\ 0 \end{pmatrix},$$

$$\begin{pmatrix} 1, & -1 \\ 1, & -1+10^{-5} \end{pmatrix} \begin{pmatrix} x_1^2 \\ x_2^2 \end{pmatrix} = \begin{pmatrix} 0 \\ 1 \end{pmatrix},$$

obtaining

$$\mathbf{x}^1 = \begin{pmatrix} 1-10^5 \\ -10^5 \end{pmatrix}, \qquad \mathbf{x}^2 = \begin{pmatrix} 10^5 \\ 10^5 \end{pmatrix}.$$

So

$$X = \begin{pmatrix} 1-10^5, & 10^5 \\ -10^5, & 10^5 \end{pmatrix} = A^{-1}.$$

15. In case (i), $\mathbf{z}$ is obviously quite inaccurate. In case (ii), $\mathbf{z}$ may be accurate but $\mathbf{x} = A^{-1}\mathbf{b}$ shows that A^{-1} must have some "large" elements. This means the problem is poorly posed.

16. Band matrices arise in many sources. The matrix defining a cubic spline has a band width of 3 and is usually called tridiagonal. We devised a special algorithm for it in II.1.3, which amounts to elimination without scaling or pivoting, as it is known to be unnecessary in this special case. The modified LASSOL would work nearly as well, even though it was designed for more general problems.

The point of the modification is to see whether the unknown to be eliminated does not appear in the equation, and so keep from doing a lot of unnecessary operations. For a non-singular triangular matrix, no eliminations are to be made but the general purpose code "eliminates" all the unknowns anyway.

Part V

PROGRAMS

In this chapter we present programs implementing the methods of the preceding chapters. In each case we describe how to use the code and present an example. The codes contain many comments to make the flow of computation easy to follow. Because they are relatively short and are more or less straightforward implementations of the algorithms described in the text, we have not flow-charted them. We urge the reader to do so, however. The codes have been provided to relieve the reader of the burden of writing them himself, but he should certainly understand what we have done. The reader might like to rewrite them to suit his particular needs or computing facility. We believe them to be adequate for class use at typical university computing installations using FORTRAN.

A number of the codes require the unit round-off error U. This is the largest number such that $1.0 \oplus U = 1.0$. Only a rough value for U is necessary in the codes, but a value appropriate to the machine being used can be important. It is certainly accurate enough simply to find U in the form $m \times 10^{-n}$ where $1 \leq m \leq 9$; this is easy enough to do in FORTRAN without examining the machine structure in detail. For example, this leads to

$$U = 9 \times 10^{-7} \text{ for the IBM S/360,}$$

$$U = 7 \times 10^{-9} \text{ for the PDP-10,}$$

$$U = 7 \times 10^{-15} \text{ for the CDC 6600.}$$

All examples were run on an IBM 360/67 computer using the WATFOR compiler. When running under WATFOR, the function name DBLE used in the codes LASSOL and LEAST must appear in a double precision statement. Since this is not common in FORTRAN compilers, the name DBLE does not appear in the double precision type statements of these two program listings.

POLYNOMIAL INTERPOLATION

The subroutine INTPOL evaluates the interpolating polynomial using divided differences, as described in II.1.1. The call list is

INTPOL(N, XN, FN, X, ABSERR, MAXDEG, ANS, ERROR)

N is the number of nodes, and must be at least 2. The lowest degree polynomial calculated is zero, which requires one node; the extra node is for the error estimate. XN and FN are the arrays of nodes and corresponding function values; they must be dimensioned in the calling program. X is the point at which the approximating polynomial is to be evaluated. The code attempts to produce a result accurate to within a specified absolute error ABSERR. The degree of the interpolating polynomial is increased until the magnitude of the estimated error ERROR is no more than ABSERR, if this is possible. ERROR is formed as described in II.1.1, and for reasonably smooth functions is a good approximation to the actual error; it is returned as output. MAXDEG is the maximum degree of interpolating polynomial to be permitted. If the above error tolerance is met, the degree of the polynomial actually used is returned in MAXDEG. Otherwise, MAXDEG remains as on input. Since MAXDEG is also used for output, it must be a variable in the calling program. To force the code to use a particular degree MAXDEG, set ABSERR negative on input. MAXDEG must be less than or equal to N − 2; if it exceeds this value, it is set equal to N − 2. The value of the interpolating polynomial is returned in ANS.

The code internally reorders the nodes by sorting their indices. This is done so that nodes as close as possible to the point X will be used. As suggested in II.1.1, this is a valuable practical device for improving accuracy in interpolation. If interpolation is to be done at more than one point X, the routine must be called for each such point.

Suppose one wants to approximate exp(0.2345) from the data

x	e^x
0.0	1.000000
0.2	1.221403
0.4	1.491825
0.6	1.822119
0.8	2.225541
1.0	2.718282

The following program gave the results indicated:

```
DIMENSION XN(6), FN(6)
XN(1) = 0.0
```

```
      XN(2) = 0.2
      XN(3) = 0.4
      XN(4) = 0.6
      XN(5) = 0.8
      XN(6) = 1.0
      FN(1) = 1.000000
      FN(2) = 1.221403
      FN(3) = 1.491825
      FN(4) = 1.822119
      FN(5) = 2.225541
      FN(6) = 2.718282
      X = 0.2345
      N = 6
      ABSERR = 5.0E-5
      MAXDEG = N - 2
      CALL INTPOL(N,XN,FN,X,ABSERR,MAXDEG,ANS,ERROR)
      PRINT 1, X, ANS, ERROR, MAXDEG
    1 FORMAT(F6.4,1P2E 17.7,I4)
      STOP
      END

      .2345     1.264248E+00     3.065744E-05     3
```

Thus, the approximation 1.264248 was obtained using a third degree polynomial with an estimated error of about 3.1×10^{-5}. The true value is 1.264276.

```
      SUBROUTINE INTPOL(N,XN,FN,X,ABSERR,MAXDEG,ANS,ERROR)
C
C  INTPOL CONSTRUCTS AN INTERPOLATING POLYNOMIAL USING
C  DIVIDED DIFFERENCES.  THE USER CAN EITHER SPECIFY THE
C  DEGREE TO BE USED OR A TOLERANCE AND A MAXIMUM DEGREE.
C  IN THE LATTER CASE THE CODE USES THE LOWEST DEGREE POLY-
C  NOMIAL WHICH IT BELIEVES MEETS THE TOLERANCE.
C
C       N - NUMBER OF NODES.  N MUST BE AT LEAST 2.  THE CODE
C          DOES NOT TEST FOR THIS.
C       XN - ARRAY OF NODES, MUST BE DISTINCT.  THE CODE DOES
C          NOT TEST FOR THIS.
C       FN - ARRAY OF FUNCTION VALUES CORRESPONDING TO NODES XN.
C       X - POINT AT WHICH INTERPOLATING POLYNOMIAL IS TO BE
C          EVALUATED.
C       ANS  - VALUE OF THE INTERPOLATING POLYNOMIAL AT X.
C       ERROR  - ESTIMATED ERROR OF ANS.  THE VALUE ANS+ERROR
C          IS OFTEN A MORE ACCURATE RESULT BUT NOT ALWAYS.
C       ABSERR - THE CODE TRIES TO CHOOSE THE DEGREE OF THE
C          INTERPOLATING POLYNOMIAL SO THAT ABS(ERROR).LE.ABSERR.
C          TO SPECIFY THE DEGREE, SET ABSERR NEGATIVE AND USE MAXDEG
C          AS DESCRIBED BELOW.
C       MAXDEG - UPPER BOUND ON THE DEGREE OF THE INTERPOLATING
C          POLYNOMIAL.  IF TOLERANCE IS MET, THE DEGREE OF THE
```

```
C           POLYNOMIAL USED IS RETURNED IN MAXDEG.  OTHERWISE,
C           MAXDEG REMAINS AS ON INPUT AND THE ERROR TOLERANCE MAY
C           NOT HAVE BEEN MET.  IN THIS CASE THE USER SHOULD CHECK
C           THE OUTPUT QUANTITY ERROR.  IF A POLYNOMIAL OF SPECIFIED
C           DEGREE IS DESIRED, SET MAXDEG TO THE DESIRED DEGREE AND
C           ABSERR TO ANY NEGATIVE VALUE.  MAXDEG MUST BE LESS THAN
C           OR EQUAL TO N-2 (FOR DEGREE K A TOTAL OF K+2 POINTS ARE
C           REQUIRED TO EVALUATE THE POLYNOMIAL AND TO ESTIMATE THE
C           ERROR). IF IT EXCEEDS THIS VALUE, IT IS SET EQUAL TO N-2.
C           SINCE MAXDEG IS USED FOR OUTPUT AS WELL AS INPUT, IT MUST
C           BE A VARIABLE IN THE CALLING PROGRAM.
C
C  NOTE THAT THE ARRAYS XN AND FN MUST BE DIMENSIONED IN THE
C  CALLING PROGRAM.
C
      DIMENSION XN(N),FN(N),
C
C  INTPOL IS WRITTEN TO HANDLE PROBLEMS WITH UP TO 10 NODES.
C  IF MORE NODES ARE DESIRED, ONLY THE NEXT STATEMENT NEED BE
C  CHANGED.  THE DIMENSION OF THE WORKING STORAGE ARRAYS V
C  AND INDEX SHOULD BE (L,L) AND (L) RESPECTIVELY WHERE L.GE.N.
C
     1 V(10,10),INDEX(10)
      MAXDEG=MIN0(MAXDEG,N-2)
      L=MAXDEG+2
      LIMIT=MIN0(L,N-1)
C
C  DETERMINE AN ORDER FOR THE NODES XN(I) (STORED INTO THE
C  ARRAY INDEX) SUCH THAT XN(INDEX(1)) IS THE NODE CLOSEST
C  TO X, XN(INDEX(2)) IS THE SECOND CLOSEST, ETC.  THE ARRAY
C  XN IS NOT ALTERED.
C
      DO 1 I=1,N
      V(I,1)=ABS(XN(I)-X)
    1 INDEX(I)=I
      DO 3 I=1,LIMIT
      IP1=I+1
      DO 2 J=IP1,N
      II=INDEX(I)
      IJ=INDEX(J)
      IF(V(II,1).LE.V(IJ,1))GO TO 2
      ITEMP=INDEX(I)
      INDEX(I)=INDEX(J)
      INDEX(J)=ITEMP
    2 CONTINUE
    3 CONTINUE
C
C  EVALUATE INTERPOLATING POLYNOMIAL AT X.
C
      PROD=1.0
      I1=INDEX(1)
      ANS=FN(I1)
      V(1,1)=FN(I1)
      DO 5 K=2,L
      IK=INDEX(K)
      V(K,1)=FN(IK)
      KM1=K-1
      DO 4 I=1,KM1
      II=INDEX(I)
    4 V(K,I+1)=(V(I,I)-V(K,I))/(XN(II)-XN(IK))
      IKM1=INDEX(KM1)
      PROD=(X-XN(IKM1))*PROD
      ERROR=PROD*V(K,K)
      IF(ABS(ERROR).GT.ABSERR)GO TO 5
      MAXDEG=K-2
      RETURN
    5 ANS=ANS+ERROR
      ANS=ANS-ERROR
      RETURN
      END
```

*LEAST SQUARES APPROXIMATION

A least squares polynomial approximation is computed using the algorithm of II.1.2 in the subroutine LEAST. It can be evaluated by the function subprogram EVAL.

The call list of LEAST is

LEAST(M, X, F, W, EPS, MAXDEG, NDEG, ARRAY, R)

There are M data points stored in the array X of the independent variables and the array F of corresponding dependent variables. A set of weights is specified in the array W; each weight must be positive, i.e., W(I) > 0 for I = 1,. . . , M. On input, EPS is the desired weighted RMS error. The code increases the degree of the fit in an attempt to meet this error request. The quantity MAXDEG is the highest degree fit allowed. The user must make MAXDEG $\leq$ M − 1. As the degree of the fit is increased, the code monitors the reduction in the error. It can happen that the error increases as the degree is raised. This is a sure sign that accuracy is being limited by the machine arithmetic, so the code returns with the fit giving the smallest error. Thus, there are three reasons for returning: the maximum degree is reached, the error started to increase, or the tolerance was met. On return, EPS is set to the weighted RMS error of the fit used and NDEG is the degree of the fit. Because EPS is used for both input and output, it must be a variable in the calling program. To force the code to use a particular degree MAXDEG, set EPS negative on input. The double precision vector R of M words outputs the double precision values of the polynomial fit at each of the data points X(I).

The vector ARRAY must provide at least 2M + 3MAXDEG words of storage. This array specifies the orthogonal polynomial fit and provides working storage. Since one may want to use many data points, the user is required to supply working storage in this program. Locations 1, . . . , MAXDEG of ARRAY hold the coefficients $a_1, \ldots, a_{\text{MAXDEG}}$ defining the orthogonal polynomials. Similarly, locations IB, . . . , IB + MAXDEG − 2 where IB = MAXDEG + 1 hold $b_2, \ldots, b_{\text{MAXDEG}}$; and locations IC, . . . , IC + MAXDEG where IC = IB + MAXDEG − 1 hold $c_0, \ldots, c_{\text{MAXDEG}}$. Storage for two orthogonal polynomials evaluated at each X(I) follow, requiring 2M locations. They are stored consecutively beginning at locations I0 and I1. The definitions of I0 and I1 are internally swapped as overwriting takes place.

The function EVAL has the call list

EVAL(Y, N, ARRAY, MAXDEG)

It evaluates the orthogonal polynomial fit computed by LEAST and specified by the vector ARRAY from that program. The fit of degree N is evaluated at the argument Y. The degree N does not have to be the degree NDEG returned by LEAST because of the properties of fits by orthogonal polynomials, but the user must be sure that N $\leq$ NDEG. LEAST is called only once for each fit, but EVAL is called once for each argument at which we require the value of the fit. Note that LEAST itself returns double precision values in the array R of the fit at the data points. By fitting a polynomial of degree, say, 5 one can use EVAL to obtain the values of the best fits of all degrees 0, 1, . . . , 5 by varying N; this can be quite useful.

To illustrate the use of these codes we evaluate the least squares cubic polynomial fit to the data given in the interpolation example at the beginning of this chapter at $x = 0.2345$. The following program gave the results indicated.

```
      DIMENSION X(6), F(6), W(6), ARRAY (15)
      DOUBLE PRECISION R(6)
      DO 1 I = 1, 6
      X(I) = 0.2* FLOAT(I − 1)
    1 W(I) = 1.0
      F(1) = 1.0
      F(2) = 1.221403
      F(3) = 1.491825
      F(4) = 1.822119
      F(5) = 2.225541
      F(6) = 2.718282
      EPS = −1.0
      CALL LEAST (6, X, F, W, EPS, 3, NDEG, ARRAY, R)
      ANS = EVAL (0.2345, 3, ARRAY, 3)
      PRINT 2, NDEG, EPS, ANS
    2 FORMAT (I15, 1P2E15.7)
      STOP
      END

      3     4.119453E−04     1.264697E+00
```

The approximation 1.264697 was obtained using a third degree polynomial fit with an RMS error of about 4.1×10^{-4}. The true value is 1.264276.

```
      SUBROUTINE LEAST(M,X,F,W,EPS,MAXDEG,NDEG,ARRAY,R)
C
C  THE SUBROUTINE LEAST AND THE FUNCTION EVAL CALCULATE
C  THE LEAST SQUARES POLYNOMIAL APPROXIMATION TO A SET
C  OF DATA SPECIFIED BY THE ARRAY OF M NODES, X, WITH CORRES-
C  PONDING FUNCTION VALUES AND WEIGHTS IN THE ARRAYS F AND W,
C  RESPECTIVELY.  THE WEIGHTS MUST ALL BE POSITIVE.  THE
C  POLYNOMIAL IS DETERMINED IN LEAST AND EVALUATED IN EVAL.
C  ON INPUT EPS IS THE DESIRED WEIGHTED RMS ERROR.  THE
C  CODE INCREASES THE DEGREE OF THE FIT IN AN ATTEMPT TO
C  TO MEET THIS ERROR REQUEST.  ON RETURN EPS IS SET TO
C  THE WEIGHTED RMS ERROR OF THE FIT.  BECAUSE EPS IS USED
C  FOR BOTH INPUT AND OUTPUT, IT MUST BE A VARIABLE IN
C  IN THE CALLING PROGRAM.  MAXDEG IS THE HIGHEST DEGREE
```

```
C   ALLOWED AND MUST BE LESS THAN OR EQUAL TC (M-1).  THE
C   ACTUAL DEGREE OF THE FIT IS RETURNED IN NDEG.  TO FORCE
C   THE CODE TO USE THE PARTICULAR DEGREE MAXDEG, SET
C   EPS NEGATIVE ON INPUT.  THE DCUBLE PRECISION VECTOR
C   R OF M WORDS OUTPUTS THE DOUBLE PRECISION VALUES OF
C   THE PCLYNCMIAL FIT AT EACH OF THE DATA POINTS X(I).
C   THE VECTOR ''ARRAY'' SPECIFIES THE ORTHOGONAL POLYNOMIAL
C   FIT AND PROVIDES WORKING STORAGE.  THE DIMENSION OF
C   ARRAY IN THE CALLING PROGRAM MUST BE AT LEAST 2*M+3*MAXDEG.
C   THE ARRAYS X, F, W, ARRAY AND R MUST BE DIMENSIONED IN
C   THE CALLING PROGRAM.
C
      DIMENSION X(1),F(1),W(1),ARRAY(1)
      DOUBLE PRECISION R(1),SUM,CK,TEMP
C
C   INITIALIZE STORAGE AND CONSTANTS.
C
      IB=MAXDEG+1
      IBL2=MAXDEG-1
      IC=IB+IBL2
      I0L1=IC+MAXDEG
      I1L1=I0L1+M
      RM=M
      TOL=RM*EPS**2
C
C   CALCULATE CONSTANT FIT.
C
      NDEG=0
      S=C.0
      SUM=0.0D0
      DO 1 I=1,M
      S=S+W(I)
    1 SUM=SUM+DBLE(W(I))*DBLE(F(I))
      RNO=S
C
C   CK  IS THE COEFFICIENT C(0) HERE.
C
      CK=SUM/RNO
      ARRAY(IC)=CK
      ERROR=0.0
      DO 2 I=1,M
      R(I)=CK
    2 ERROR=ERROR+W(I)*SNGL(CK-DBLE(F(I)))**2
      IF(NDEG.EQ.MAXDEG)GO TO 14
      IF(EPS.LT.0.0)GO TO 3
      IF(ERROR.LE.TOL)GO TO 14
C
C   CALCULATE LINEAR FIT.
C
    3 NDEG=1
      ES=ERROR
      SUM=0.0D0
      DO 4 I=1,M
    4 SUM=SUM+DBLE(W(I))*DBLE(X(I))
C
C   CALCULATE A(1).
C
      ARRAY(1)=SUM/RNO
C
C   CALCULATE Q1(.).
C
      S=0.0
      SUM=0.0D0
      DO 5 I=1,M
      ARRAY(I1L1+I)=X(I)-ARRAY(1)
      S=S+W(I)*ARRAY(I1L1+I)**2
      TEMP=DBLE(F(I))-R(I)
    5 SUM=SUM+DBLE(W(I))*DBLE(ARRAY(I1L1+I))*TEMP
      RN1=S
C
```

```
C  CK   IS THE COEFFICIENT C(1) HERE.
C
      CK=SUM/RN1
      ARRAY(IC+1)=CK
C
C  CALCULATE THE VALUE OF THE FIT AT THE DATA POINTS AND
C  ALSO THE RMS ERROR.
C
      ERROR=0.0
      DO 6 I=1,M
      R(I)=R(I)+CK*DBLE(ARRAY(I1L1+I))
    6 ERROR=ERROR+W(I)*SNGL(R(I)-DBLE(F(I)))**2
      IF(ERROR.GT.ES.AND.EPS.GE.0.0)GO TO 12
      IF(NDEG.EQ.MAXDEG)GO TO 14
      IF(ERROR.LE.TOL.AND.EPS.GE.0.0)GO TO 14
      DO 7 I=1,M
    7 ARRAY(IOL1+I)=1.0
      NDEG=2
      K=2
C
C  GENERAL FIT.
C
    8 ES=ERROR
C
C  CALCULATE B(K).
C
      ARRAY(IBL2+K)=RN1/RN0
C
C  CALCULATE A(K).
C
      SUM=0.0D0
      DO 9 I=1,M
    9 SUM=SUM+DBLE(W(I))*DBLE(X(I))*DBLE(ARRAY(I1L1+I))**2
      ARRAY(K)=SUM/RN1
C
C  CALCULATE QK(.) OVERWRITING ON QK-2(.).
C
      S=0.0
      SUM=0.0D0
      DO 10 I=1,M
      ARRAY(IOL1+I)=(X(I)-ARRAY(K))*ARRAY(I1L1+I)
     1 -ARRAY(IBL2+K)*ARRAY(IOL1+I)
      S=S+W(I)*ARRAY(IOL1+I)**2
      TEMP=DBLE(F(I))-R(I)
   10 SUM=SUM+DBLE(W(I))*DBLE(ARRAY(IOL1+I))*TEMP
      RN0=RN1
      RN1=S
C
C  SWAP INDICES SO I1 REFERS TO STORAGE OF QK(.)
C  AND IO TO QK-1(.).
C
      IT=IOL1
      IOL1=I1L1
      I1L1=IT
C
C  CK IS THE COEFFICIENT C(K) HERE.
C
      CK=SUM/RN1
      ARRAY(IC+K)=CK
C
C  CALCULATE THE VALUE OF THE FIT AT THE DATA POINTS AND
C  ALSO THE RMS ERROR.
C
      ERROR=0.0
      DO 11 I=1,M
      R(I)=R(I)+CK*DBLE(ARRAY(I1L1+I))
   11 ERROR=ERROR+W(I)*SNGL(R(I)-DBLE(F(I)))**2
      IF(ERROR.GT.ES.AND.EPS.GE.0.0)GO TO 12
      IF(NDEG.EQ.MAXDEG)GO TO 14
      IF(ERROR.LE.TOL.AND.EPS.GE.0.0)GO TO 14
```

```
      NDEG=NDEG+1
      K=K+1
      GC TO 8
C
C  HERE IF ERROR INCREASED ON RAISING DEGREE.
C
 12   NDEG=NDEG-1
      ERROR=ES
      DO 13 I=1,M
 13   R(I)=R(I)-CK*DBLE(ARRAY(I1L1+I))
C
C  EXIT.
C
 14   EPS=SQRT(ERROR/RM)
      RETURN
      END
      FUNCTION EVAL(Y,N,ARRAY,MAXDEG)
C
C  THE FUNCTION EVAL EVALUATES THE ORTHOGONAL POLYNOMIAL
C  FIT COMPUTED BY LEAST AND SPECIFIED BY THE VECTOR
C  ARRAY.  THE FIT OF DEGREE N IS EVALUATED AT THE ARGU-
C  MENT Y.  N MUST BE LESS THAN OR EQUAL TO NDEG AS
C  RETURNED FROM LEAST.  LEAST IS CALLED ONLY ONCE FOR
C  EACH FIT, BUT EVAL IS CALLED ONCE FOR EACH ARGUMENT
C  AT WHICH WE REQUIRE THE VALUE OF THE FIT.  MAXDEG MUST
C  HAVE THE SAME VALUE AS IN THE CALL TO LEAST.
C
      DIMENSION ARRAY(1)
      IB=MAXDEG+1
      IC=MAXDEG+IB-1
C
C  EVALUATE N=0,1 AS SPECIAL CASES.
C
      IF(N.GT.0)GO TO 1
      EVAL=ARRAY(IC)
      RETURN
  1   IF(N.GT.1)GO TO 2
      EVAL=ARRAY(IC)+ARRAY(IC+1)*(Y-ARRAY(1))
      RETURN
C
C  GENERAL RECURRENCE RELATION
C
  2   DKP2=ARRAY(IC+N)
      DKP1=ARRAY(IC+N-1)+(Y-ARRAY(N))*DKP2
      NL2=N-2
      IF(NL2.LT.1)GO TO 4
      DO 3 L=1,NL2
      K=1+NL2-L
      DK=ARRAY(IC+K)+(Y-ARRAY(K+1))*DKP1
     1  -ARRAY(IB+K)*DKP2
      DKP2=DKP1
  3   DKP1=DK
  4   EVAL=ARRAY(IC)+(Y-ARRAY(1))*DKP1
     1  -ARRAY(IB)*DKP2
      RETURN
      END
```

*CUBIC INTERPOLATORY SPLINES

The subroutine SPCOEF and the function SPLINE are designed to evaluate the natural cubic interpolatory spline fit to a set of data specified by the array of N nodes, XN, with corresponding function values in the array FN. The nodes of XN must be distinct. The spline is determined in SPCOEF and is evaluated in SPLINE. SPCOEF arranges the nodes in increasing order and stores this order in the array INDEX. Then XN(INDEX(1)) is the smallest node, XN(INDEX(2)) the next largest, etc. The array XN itself is not altered. SPCOEF then evaluates the array S of the N second derivatives needed to define the spline. The call arguments of the subroutine subprogram SPCOEF are

SPCOEF(N, XN, FN, S, INDEX)

The remainder of the spline calculation is done by the function subprogram SPLINE, which has the call arguments

SPLINE(N, XN, FN, S, INDEX, X)

Here N, XN, FN, S, and INDEX are as before, and X is the point at which we wish to evaluate the spline. Its value is returned by the function name SPLINE. The arrays XN, FN, and INDEX must be dimensioned in the calling program.

If X < XN(INDEX(1)), the function is approximated by a straight line which passes through the point (XN(INDEX(1)), FN(INDEX(1))) with a slope which agrees with the slope of the spline at that point. If X > XN(INDEX(N)), a straight line approximation is again used; this time it passes through (XN(INDEX(N)), FN(INDEX(N))) and has the slope of the spline at that point.

If we want to approximate a function $f(x)$ by a spline at several points x given a set of nodes XN, the routine SPCOEF need be called only once, and SPLINE is called once for each point x. This will be illustrated in the example to follow.

Suppose we have the data

x	$f(x)$
0.0	0.0000
0.2	1.2214
0.4	1.4918
0.6	1.8221
0.8	2.2255
1.0	2.7183

and we wish to approximate $f(x)$ by the value of a cubic spline at $x = 0.1$. We could use the following program.

```
      DIMENSION XN(6), FN(6), S(6), INDEX(6)
      XN(1) = 0.0
      XN(2) = 0.2
      XN(3) = 0.4
      XN(4) = 0.6
      XN(5) = 0.8
      XN(6) = 1.0
      FN(1) = 0.0000
      FN(2) = 1.2214
      FN(3) = 1.4918
      FN(4) = 1.8221
      FN(5) = 2.2255
      FN(6) = 2.7183
      N = 6
      CALL SPCOEF (N, XN, FN, S, INDEX)
      X = 0.1
      SP = SPLINE (N, XN, FN, S, INDEX, X)
      PRINT 10, X, SP
   10 FORMAT (1X, F6.2, 1PE16.7)
      STOP
      END
```

If we had wanted to generate the approximations for, say, $x = 0.1, 0.3, \ldots,$ 0.9, we could modify the program to read:

```
        .
        .
        .
      CALL SPCOEF (N, XN, FN, S, INDEX)
      DO 1 I = 1, 5
      X = FLOAT (2*I - 1)*0.1
      SP = SPLINE (N, XN, FN, S, INDEX, X)
    1 PRINT 10, X, SP
        .
        .
        .
```

The output for the first program is

```
   0.10    1.1070240E+00
```

and for the second

```
   0.10    1.1070240E+00
   0.30    1.3492510E+00
   0.50    1.6491810E+00
   0.70    2.0123460E+00
   0.90    2.4645410E+00.
```

```
      SUBROUTINE SPCOFF(N,XN,FN,S,INDEX)
C
C  THE SUBROUTINE SPCOEF AND THE FUNCTION SPLINE CALCULATE THE
C  NATURAL CUBIC INTERPOLATORY SPLINE FIT TO THE DATA SPECIFIED
C  BY THE ARRAY OF N NODES XN, WITH CORRESPONDING FUNCTION
C  VALUES IN THE ARRAY FN.  THE NODES XN MUST BE DISTINCT.  THE
C  SPLINE IS DETERMINED IN SPCOEF AND EVALUATED IN SPLINE.  SPCOEF
C  ARRANGES THE NODES IN INCREASING ORDER AND STORES THIS ORDER IN
C  THE ARRAY INDEX.  THE ARRAY XN ITSELF IS NOT ALTERED.  SPCOEF
C  THEN CALCULATES THE ARRAY S OF SECOND DERIVATIVES NEEDED TO
C  DEFINE THE SPLINE.  THE ARRAYS XN, FN, S, AND INDEX MUST BE
C  DIMENSIONED IN THE CALLING PROGRAM.
C
      DIMENSION XN(N),FN(N),S(N),INDEX(N),
C
C  SPCOEF IS WRITTEN TO HANDLE PROBLEMS WITH UP TO TWENTY FIVE
C  NODES.  IF MORE NODES ARE USED, ONLY THE NEXT STATEMENT
C  NEED BE CHANGED.  THE DIMENSION OF THE ARRAYS RHO AND TAU
C  MUST BE AT LEAST N.
C
     1 RHO(25),TAU(25)
      NM1=N-1
C
C  ARRANGE THE NODES XN IN INCREASING ORDER.  STORE THE
C  ORDER IN THE ARRAY INDEX.
C
      DO 1 I=1,N
    1 INDEX(I)=I
      DO 3 I=1,NM1
      IP1=I+1
      DO 2 J=IP1,N
      II=INDEX(I)
      IJ=INDEX(J)
      IF(XN(II).LE.XN(IJ))GO TO 2
      ITEMP=INDEX(I)
      INDEX(I)=INDEX(J)
      INDEX(J)=ITEMP
    2 CONTINUE
    3 CONTINUE
      NM2=N-2
C
C  CALCULATE THE ELEMENTS OF THE ARRAYS RHO AND TAU.
C
      RHO(2)=0.0
      TAU(2)=0.0
      DO 4 I=2,NM1
      IIM1=INDEX(I-1)
      II=INDEX(I)
      IIP1=INDEX(I+1)
      HIM1=XN(II)-XN(IIM1)
      HI=XN(IIP1)-XN(II)
      TEMP=(HIM1/HI)*(RHO(I)+2.0)+2.0
      RHO(I+1)=-1.0/TEMP
      D=6.0*((FN(IIP1)-FN(II))/HI-(FN(II)-FN(IIM1))/HIM1)/HI
    4 TAU(I+1)=(D-HIM1*TAU(I)/HI)/TEMP
C
C  COMPUTE ARRAY OF SECOND DERIVATIVES S FOR THE NATURAL SPLINE.
C
      S(1)=0.0
      S(N)=0.0
      DO 5 I=1,NM2
      IB=N-I
    5 S(IB)=RHO(IB+1)*S(IB+1)+TAU(IB+1)
      RETURN
      END
```

```
      FUNCTION SPLINE(N,XN,FN,S,INDEX,X)
C
C  THE FUNCTION SPLINE ACCEPTS AS INPUT THE QUANTITIES N, XN, FN,
C  S, AND INDEX AS DEFINED IN THE SUBROUTINE SPCOFF AND A NUMBER
C  X AT WHICH THE SPLINE IS TO BE EVALUATED.  SPCOEF IS CALLED
C  ONCE FOR EACH FIT, BUT SPLINE IS CALLED ONCE FOR EACH ARGUMENT
C  AT WHICH WE REQUIRE THE VALUE OF THE FIT.
C
      DIMENSION XN(N),FN(N),S(N),INDEX(N)
C
C  IF X.LT.XN(INDEX(1)), APPROXIMATE FUNCTION BY THE STRAIGHT
C  LINE WHICH PASSES THROUGH THE POINT (XN(INDEX(1)),FN(INDEX(1)))
C  AND WHOSE SLOPE AGREES WITH THE SLOPE OF THE SPLINE AT THAT POINT.
C
      I1=INDEX(1)
      IF(X.GE.XN(I1))GO TO 1
      I2=INDEX(2)
      H1=XN(I2)-XN(I1)
      SPLINE=FN(I1)+(X-XN(I1))*((FN(I2)-FN(I1))/H1-H1*S(2)/6.0)
      RETURN
C
C  IF X.GE.XN(INDEX(N)), APPROXIMATE FUNCTION BY THE STRAIGHT LINE
C  WHICH PASSES THROUGH THE POINT (XN(INDEX(N)),FN(INDEX(N))) AND
C  WHOSE SLOPE AGREES WITH THE SLOPE OF THE SPLINE AT THAT POINT.
C
    1 IN=INDEX(N)
      IF(X.LE.XN(IN))GO TO 2
      INM1=INDEX(N-1)
      HNM1=XN(IN)-XN(INM1)
      SPLINE=FN(IN)+(X-XN(IN))*((FN(IN)-FN(INM1))/HNM1+HNM1*S(N-1)/6.
      RETURN
C
C  FOR XN(INDEX(1)).LE.X.LE.XN(INDEX(N)) CALCULATE SPLINE FIT.
C
    2 DO 3 I=2,N
      II=INDEX(I)
      IF(X.LE.XN(II))GO TO 4
    3 CONTINUE
    4 L=I-1
      IL=INDEX(L)
      ILP1=INDEX(L+1)
      A=XN(ILP1)-X
      B=X-XN(IL)
      HL=XN(ILP1)-XN(IL)
      SPLINE=A*S(L)*(A**2/HL-HL)/6.0+B*S(L+1)*(B**2/HL-HL)/6.0
     1 +(A*FN(IL)+B*FN(ILP1))/HL
      RETURN
      END
```

INTEGRATION

The subroutine SIMP is an adaptive iterative code based on Simpson's rule as described in II.2. Its call list is

SIMP(F, A, B, ACC, ANS, ERROR, AREA, IFLAG)

This subroutine attempts to integrate

$$\int_A^B F(x)\,dx$$

and return the value as ANS. F is the name of a FUNCTION subprogram which must be declared in an EXTERNAL statement in the calling program. The code attempts to produce a result accurate to a specified value ACC in the relative L_1 sense. Accordingly, it also computes the approximation

$$\text{AREA} \doteq \int_{A}^{B} |F(x)|\, dx,$$

which is used in attempting to satisfy

$$\frac{\left| \text{ANS} - \int_{A}^{B} F(x)\, dx \right|}{\int_{A}^{B} |F(x)|\, dx} \leq \text{ACC}.$$

The code estimates the error in ANS to be ERROR. So, after the fact, one can estimate the error in ANS in any of the usual ways in his calling program. For absolute error, relative error, and relative L_1 error respectively, he could estimate by

$$|\text{ERROR}|,$$
$$|\text{ERROR}|/|\text{ANS}|,$$
$$|\text{ERROR}|/|\text{AREA}|.$$

Notice that AREA can be negative because $A > B$ is permitted. If $F(x)$ is of one sign on [A, B], one would expect the values of ANS and AREA to have the same magnitude. However, they might well differ a little since they are not computed in the same way, although they are theoretically the same. AREA is more susceptible to roundoff errors.

If it is necessary to go to 30 levels or if a subinterval becomes too small for the machine word length, the code accepts the approximation at that level and works back up. IFLAG is set to 2 to warn the user that ERROR may be somewhat unreliable. If more than 2000 function evaluations are made, rough approximations are used to complete the computation and ERROR is usually unreliable. IFLAG is then set to 3. Normal return has IFLAG equal to 1. A computed GO TO statement is a convenient way to take appropriate action depending on the value of IFLAG.

The user may prefer to extrapolate ANS and use the value

$$\text{ANS} + \text{ERROR}$$

instead. Frequently, this adds another decimal place or so of accuracy. It is most likely to be useful when requesting very accurate answers; it can be harmful, however.

The following program to evaluate

$$\int_0^1 e^x \, dx = e - 1$$

illustrates SIMP. It computes the integral and compares the estimated accuracy to the true accuracy in the relative L_1 sense.

```
      EXTERNAL F
      CALL SIMP (F, 0.0, 1.0, 1.0E-5, ANS, ERROR, AREA, IFLAG)
      TRUANS = EXP(1.0) - 1.0
      TRUACC = ABS((TRUANS - ANS)/TRUANS)
      ESTACC = ABS(ERROR/AREA)
      PRINT 1, ANS, TRUANS
      PRINT 1, ESTACC, TRUACC
    1 FORMAT(1P2E20.7)
      STOP
      END
      FUNCTION F(X)
      F = EXP(X)
      RETURN
      END
```

The results printed are the computed answer, the true answer, the estimated relative L_1 accuracy, and the true relative L_1 accuracy. Note that the requested relative L_1 accuracy was 1.0E−5 and that relative L_1 is the same as relative error for this problem. The results were:

1.718281E+00   1.718282E+00

9.250289E−08   5.550162E−07.

```
      SUBROUTINE SIMP(F,A,B,ACC,ANS,ERROR,AREA,IFLAG)
C
C  SIMP IS AN ADAPTIVE, ITERATIVE CODE BASED ON SIMPSON'S RULE.
C  IT IS DESIGNED TO EVALUATE THE DEFINITE INTEGRAL OF A CON-
C  TINUOUS FUNCTION WITH FINITE LIMITS OF INTEGRATION.
C
C
C        F - NAME OF FUNCTION WHOSE INTEGRAL IS DESIRED.  THE FUNCTION
C            NAME F MUST APPEAR IN AN EXTERNAL STATEMENT IN THE CALLING
C            PROGRAM.
C        A,B - LOWER AND UPPER LIMITS OF INTEGRATION.
C        ANS- APPROXIMATE VALUE OF THE INTEGRAL OF F(X) FROM
C            A TO B.
C        AREA - APPROXIMATE VALUE OF THE INTEGRAL OF ABS(F(X))
C            FROM A TO B.
C        ERROR - ESTIMATED ERROR OF ANS.  USER MAY WISH TO EXTRA-
C            POLATE BY FORMING ANS+ERROR TO GET WHAT IS OFTEN A
C            MORE ACCURATE RESULT, BUT NOT ALWAYS.
C        ACC - DESIRED ACCURACY OF ANS.  CODE TRIES TO MAKE
C            ABS(ERROR).LE.ACC*ABS(AREA).
```

```
C       IFLAG = 1 FOR NORMAL RETURN.
C             = 2 IF IT IS NECESSARY TO GO TO 30 LEVELS OR
C                 USE A SUBINTERVAL TOO SMALL FOR MACHINE WORD
C                 LENGTH.  ERROR MAY BE UNRELIABLE IN THIS CASE.
C             = 3 IF MORE THAT 2000 FUNCTION EVALUATIONS ARE USED.
C                 ROUGH APPROXIMATIONS ARE USED TO COMPLETE THE
C                 COMPUTATIONS AND ERROR IS USUALLY UNRELIABLE.
C
      DIMENSION FV(5),LORR(30),F1T(30),F2T(30),F3T(30),DAT(30),
     1  ARESTT(30),ESTT(30),EPST(30),PSUM(30)
C
C  SET U TO APPROXIMATELY THE UNIT ROUND-OFF OF SPECIFIC MACHINE
C  (HERE IBM 360/67)
C
      U = 9.0E-7
C
C  INITIALIZE
C
      FOURU=4.0*U
      IFLAG=1
      EPS=ACC
      ERROR=0.0
      LVL=1
      LORR(LVL)=1
      PSUM(LVL)=0.0
      ALPHA=A
      DA=B-A
      AREA=0.0
      AREST=0.0
      FV(1)=F(ALPHA)
      FV(3)=F(ALPHA+0.5*DA)
      FV(5)=F(ALPHA+DA)
      KOUNT=3
      WT=DA/6.0
      EST=WT*(FV(1)+4.0*FV(3)+FV(5))
C
C  'BASIC STEP'.  HAVE ESTIMATE EST OF INTEGRAL ON (ALPHA,ALPHA+DA).
C  BISECT AND COMPUTE ESTIMATES ON LEFT AND RIGHT HALF INTERVALS.
C  SIMILARLY TREAT INTEGRAL OF ABS(F(X)).  SUM IS BETTER VALUE FOR
C  INTEGRAL AND DIFF/15.0 IS APPROXIMATELY ITS ERROR.
C
    1 DX=0.5*DA
      FV(2)=F(ALPHA+0.5*DX)
      FV(4)=F(ALPHA+1.5*DX)
      KOUNT=KOUNT+2
      WT=DX/6.0
      ESTL=WT*(FV(1)+4.0*FV(2)+FV(3))
      ESTR=WT*(FV(3)+4.0*FV(4)+FV(5))
      SUM=ESTL+ESTR
      ARESTL=WT*(ABS(FV(1))+ABS(4.0*FV(2))+ABS(FV(3)))
      ARESTR=WT*(ABS(FV(3))+ABS(4.0*FV(4))+ABS(FV(5)))
      AREA=AREA+((ARESTL+ARESTR)-AREST)
      DIFF=EST-SUM
C
C  IF ERROR IS ACCEPTABLE, GO TO 2.  IF INTERVAL IS TOO SMALL OR
C  TOO MANY LEVELS OR TOO MANY FUNCTION EVALUATIONS, SET A FLAG
C  AND GO TO 2 ANYWAY.
C
      IF(ABS(DIFF).LE.EPS*ABS(AREA))GO TO 2
      IF(ABS(DX).LE.FOURU*ABS(ALPHA))GO TO 5
      IF(LVL.GE.30)GO TO 5
      IF(KOUNT.GE.2000)GO TO 6
C
C  HERE TO RAISE LEVEL.  STORE INFORMATION TO PROCESS RIGHT HALF
C  INTERVAL LATER.  INITIALIZE FOR 'BASIC STEP' SO AS TO TREAT
C  LEFT HALF INTERVAL.
C
      LVL=LVL+1
      LORR(LVL)=0
      F1T(LVL)=FV(3)
```

```
      F2T(LVL)=FV(4)
      F3T(LVL)=FV(5)
      DA=DX
      DAT(LVL)=DX
      AREST=ARESTL
      ARESTT(LVL)=ARESTR
      EST=ESTL
      ESTT(LVL)=ESTR
      EPS=EPS/1.4
      EPST(LVL)=EPS
      FV(5)=FV(3)
      FV(3)=FV(2)
      GO TO 1
C
C  ACCEPT APPROXIMATE INTEGRAL SUM.  IF IT WAS ON A LEFT INTERVAL,
C  GO TO 'MOVE RIGHT'.  IF A RIGHT INTERVAL, ADD RESULTS TO FIN-
C  ISH AT THIS LEVEL.  ARRAY LORR (MNEMONIC FOR LEFT OR RIGHT)
C  TELLS WHETHER LEFT OR RIGHT INTERVAL AT EACH LEVEL.
C
    2 ERROR=ERROR+DIFF/15.0
    3 IF(LORR(LVL).EQ.0)GO TO 4
      SUM=PSUM(LVL)+SUM
      LVL=LVL-1
      IF(LVL.GT.1)GO TO 3
      ANS=SUM
      RETURN
C
C  'MOVE RIGHT'.  RESTORE SAVED INFORMATION TO PROCESS RIGHT HALF
C  INTERVAL.
C
    4 PSUM(LVL)=SUM
      LORR(LVL)=1
      ALPHA=ALPHA+DA
      DA=DAT(LVL)
      FV(1)=F1T(LVL)
      FV(3)=F2T(LVL)
      FV(5)=F3T(LVL)
      AREST=ARESTT(LVL)
      EST=ESTT(LVL)
      EPS=EPST(LVL)
      GO TO 1
C
C  ACCEPT 'POOR' VALUE.  SET APPROPRIATE FLAGS.
C
    5 IFLAG=2
      GO TO 2
    6 IFLAG=3
      GO TO 2
      END
```

ROOTS OF NONLINEAR EQUATIONS

The subroutine ZEROIN computes a root of the nonlinear equation

$$F(x) = 0$$

of a single real variable. It uses bisection and the secant rule as described in II.3. The call list is

ZEROIN(F, B, C, ABSERR, RELERR, IFLAG)

F is the name of a FUNCTION subprogram for evaluating $F(x)$; it must be declared in an EXTERNAL statement in the calling program. Normal input consists of a continuous function $F(x)$ and points B and C such that $F(B) \times F(C) < 0$; this implies that there is a root between B and C. Both B and C are also output quantities, so they must be variables in the calling program. On output it is always the case that $|F(B)| \leq |F(C)|$.

The accuracy parameters ABSERR and RELERR allow one to specify pure absolute error (RELERR = 0), pure relative error (ABSERR = 0), or a mixed test. RELERR is changed internally so that it is not less than U, the unit roundoff for the machine being used. The code attempts to bracket a root between B and C, with B being the closer approximation so that the convergence test

$$\left|\frac{B - C}{2}\right| \leq \text{RELERR} \times |B| + \text{ABSERR}$$

is satisfied. The user should be aware of the danger of a pure relative error test if zero is a possible root.

Normal output has $F(B) \times F(C) < 0$ and the convergence test met. This is signaled by IFLAG = 1. If, fortuitously, a computed function value of *exactly* zero is found, the code returns with a B such that $F(B) = 0$ but the convergence test may not be met; in particular, C need not be close to B. This has an IFLAG of 2. If the convergence test is met but $|F(B)|$ is larger than the values $|F(B)|$ and $|F(C)|$ were on input, IFLAG is set to 3. It is then likely that B is near an odd order pole of F. If the convergence test is met but $F(B) \times F(C) > 0$, there is apparently no root in the initial interval and IFLAG is set to 4. A local minimum or an approximation to an even order root may have been found, since the code seeks to make $|F(B)|$ small. If too many function evaluations are made, IFLAG is set to 5 and the code terminates before the convergence test is satisfied. In this version, 500 function evaluations are permitted. IFLAG may conveniently be used in a computed GO TO statement to take appropriate action depending on the results of the computation.

The function $F(x) = e^{-x} - x$ has $F(0) > 0$ and $F(1) < 0$; hence, the equation $F(x) = 0$ has a root between $B = 0$ and $C = 1$. The following example illustrates the use of ZEROIN.

```
      EXTERNAL F
      B = 0.0
      C = 1.0
      CALL ZEROIN (F, B, C, 0.0, 1.E-5, IFLAG)
      GO TO (1, 2, 2, 2, 3), IFLAG
    1 RESIDL = F(B)
      PRINT 100, B, RESIDL, IFLAG
  100 FORMAT (1P2E20.7, I5)
      STOP
```

```
    2  PRINT 101, B, IFLAG
  101  FORMAT (1PE20.7, I5)
       STOP
    3  PRINT 100, B, C, IFLAG
       STOP
       END
       FUNCTION F(X)
       F = EXP(-X) - X
       RETURN
       END
```

The results printed were

```
5.671433E-01     2
```

```
      SUBROUTINE ZEROIN(F,B,C,ABSERR,RELERR,IFLAG)
C
C  ZEROIN COMPUTES A ROOT OF THE NONLINEAR EQUATION F(X)=0
C  WHERE F(X) IS A CONTINUOUS REAL FUNCTION OF A SINGLE REAL
C  VARIABLE X.  THE METHOD USED IS A COMBINATION OF BISECTION
C  AND THE SECANT RULE.
C
C  NORMAL INPUT CONSISTS OF A CONTINUOUS FUNCTION F AND AN
C  INTERVAL (B,C) SUCH THAT F(B)*F(C).LE.0.0.  EACH ITERATION
C  FINDS NEW VALUES OF B AND C SUCH THAT THE INTERVAL (B,C) IS
C  SHRUNK AND F(B)*F(C).LE.0.0.  THE STOPPING CRITERION IS
C
C           ABS(B-C).LE.2.0*(RELERR*ABS(B)+ABSERR)
C
C  WHERE RELERR=RELATIVE ERROR AND ABSERR=ABSOLUTE ERROR ARE
C  INPUT QUANTITIES.  AS B AND C ARE USED FOR BOTH INPUT AND
C  OUTPUT, THEY MUST BE VARIABLES IN THE CALLING PROGRAM.
C  THE FUNCTION NAME F MUST APPEAR IN AN EXTERNAL STATEMENT IN
C  THE CALLING PROGRAM.
C
C  IF 0 IS A POSSIBLE ZERO, ONE SHOULD NOT CHOOSE ABSERR=0.0.
C
C  THE OUTPUT VALUE OF B IS THE BETTER APPROXIMATION TO A ROOT
C  AS B AND C ARE ALWAYS REDEFINED SO THAT ABS(F(B)).LE.ABS(F(C)).
C
C  A FLAG, IFLAG, IS PROVIDED AS AN OUTPUT QUANTITY.  IT MAY ASSUME
C  THE VALUES 1-5 WHERE
C
C     IFLAG=1  IF F(B)*F(C).LT.0 AND THE STOPPING CRITERION IS MET.
C
C          =2  IF A VALUE B IS FOUND SUCH THAT THE COMPUTED VALUE
C              F(B) IS EXACTLY ZERO.  THE INTERVAL (B,C) MAY NOT
C              SATISFY THE STOPPING CRITERION.
C
C          =3  IF ABS(F(B)) EXCEEDS THE INPUT VALUES ABS(F(B)),
C              ABS(F(C)).   IN THIS CASE IT IS LIKELY THAT B IS CLOSE
C              TO A POLE OF F.
C
C          =4  IF NO ODD ORDER ZERO WAS FOUND IN THE INTERVAL.  A
C              LOCAL MINIMUM MAY HAVE BEEN OBTAINED.
C
C          =5  IF TOO MANY FUNCTION EVALUATIONS WERE MADE.
C              (AS PROGRAMMED, 500 ARE ALLOWED.)
C
C
C  SET U TO APPROXIMATELY THE UNIT ROUND-OFF OF SPECIFIC MACHINE.
```

```
C   (HERE AN IBM 360/67)
C
       U=9.0E-7
C
       RE=AMAX1(RELERR,U)
       IC=0
       ACBS=ABS(B-C)
       A=C
       FA=F(A)
       FB=F(B)
       FC=FA
       KCUNT=2
       FX=AMAX1(ABS(FB),ABS(FC))
  1    IF(AES(FC).GE.ABS(FB))GO TO 2
C
C   INTERCHANGE B AND C SO THAT AES(F(B)).LE.ABS(F(C)).
C
       A=B
       FA=FB
       B=C
       FB=FC
       C=A
       FC=FA
  2    CMB=0.5*(C-B)
       ACMB=ABS(CMB)
       TOL=RE*ABS(B)+ABSERR
C
C   TEST STOPPING CRITERION AND FUNCTION COUNT.
C
       IF(ACMB.LE.TOL)GO TO 8
       IF(KOUNT.GE.500)GO TO 12
C
C   CALCULATE NEW ITERATE IMPLICITLY AS B+P/Q
C   WHERE WE ARRANGE P.GE.0.  THE IMPLICIT
C   FORM IS USED TO PREVENT OVERFLOW.
C
       P=(B-A)*FB
       Q=FA-FB
       IF(P.GE.0.0)GO TO 3
       P=-P
       Q=-Q
C
C   UPDATE A, CHECK IF REDUCTION IN THE SIZE OF BRACKETING
C   INTERVAL IS SATISFACTORY.  IF NOT, BISECT UNTIL IT IS.
C
  3    A=B
       FA=FB
       IC=IC+1
       IF(IC.LT.4)GO TO 4
       IF(8.0*ACMB.GE.ACBS)GO TO 6
       IC=0
       ACBS=ACMB
C
C   TEST FOR TOO SMALL A CHANGE.
C
  4    IF(P.GT.ABS(Q)*TOL)GO TO 5
C
C   INCREMENT BY TOLERANCE.
C
       B=B+SIGN(TOL,CMB)
       GO TO 7
C
C   ROOT CUGHT TO BE BETWEEN B AND (C+B)/2.
C
  5    IF(P.GE.CMB*Q)GO TO 6
C
C   USE SECANT RULE.
C
       B=B+P/Q
       GO TO 7
```

```
C
C  USE BISECTION.
C
  6    B=0.5*(C+B)
C
C  HAVE COMPLETED COMPUTATION FOR NEW ITERATE B.
C
  7    FB=F(B)
       IF(FB.EQ.0.0)GO TO 9
       KOUNT=KOUNT+1
       IF(SIGN(1.0,FB).NE.SIGN(1.0,FC))GO TO 1
       C=A
       FC=FA
       GO TO 1
C
C  FINISHED.  SET IFLAG.
C
  8    IF(SIGN(1.0,FB).EQ.SIGN(1.0,FC))GO TO 11
       IF(ABS(FB).GT.FX)GO TO 10
       IFLAG=1
       RETURN
  9    IFLAG=2
       RETURN
 10    IFLAG=3
       RETURN
 11    IFLAG=4
       RETURN
 12    IFLAG=5
       RETURN
       END
```

ORDINARY DIFFERENTIAL EQUATIONS

The subroutine RKF, a Runge-Kutta code based on Fehlberg's scheme, integrates a system of N first order ordinary differential equations of the form

$$Y_1' = F_1(X, Y_1, Y_2, \ldots, Y_N),$$

$$Y_2' = F_2(X, Y_1, Y_2, \ldots, Y_N),$$

$$\vdots$$

$$Y_N' = F_N(X, Y_1, Y_2, \ldots, Y_N),$$

with initial conditions given at X = A in the vector Y

$$Y_1(A) = Y(1),$$

$$Y_2(A) = Y(2),$$

$$\vdots$$

$$Y_N(A) = Y(N).$$

The code estimates its local error and adjusts its step size as described in II.4. Its call list is

RKF(A, Y, N, F, DA, H, HMX, ABSERR, RELERR, IFLAG)

F is the name of a subroutine subprogram of the form

F(X, Y, YP)

and must be declared in an EXTERNAL statement in the calling program. The components of the vector YP are defined in the subroutine F as

$$\text{YP(I)} = \text{F}_\text{I}(\text{X}, \text{Y(1)}, \text{Y(2)}, \ldots, \text{Y(N)}) \quad \text{for} \quad \text{I} = 1, 2, \ldots, \text{N}$$

where, of course, the vector Y represents the solution vector at X. In order to use variable dimensions in F, dimension Y and YP there to 1. This merely serves to tell the compiler that a singly subscripted array is being passed to F. This FORTRAN trick avoids the annoying alternative of putting N in the call list for F.

The code is to integrate from A to A + DA where DA can be negative. On input, the vector Y contains the initial values of the dependent variables; on output, it contains the solution values at the output value of A. For various reasons one may not be able to compute to A + DA, so the output value of A is the last point at which a solution was successfully computed.

The parameters ABSERR and RELERR are absolute and relative accuracy requests respectively. The code estimates the local error in the computation of each component, Y(I), *relative to the step size being used* and tests whether this estimate R(I) satisfies

$$|\text{R(I)}| \leq \text{RELERR} \times |\text{Y(I)}| + \text{ABSERR}$$

for each component. If RELERR = 0, this is a pure absolute error test; if ABSERR = 0, it is a pure relative error test; otherwise, it is a mixed test. The step size is adjusted so as to meet the accuracy request.

The quantity HMX is the maximum step size to be permitted. It should be taken small enough so that no interesting behavior of the solution will be "jumped over." Values larger than |DA| are, of course, limited to |DA| internally. An input H is the nominal step size to be tried. On output it is set to the step size the code believes suitable for continuing the integration. Because a user guess for H is often unrealistic, a forgiving approach is taken in the code; an input H is changed until it has a reasonable size and the correct sign as determined by DA.

A flag IFLAG conveys information about the code's performance. Input values of

$$\text{HMX} \leq 0, \quad \text{RELERR} < 0, \quad \text{ABSERR} < 0,$$
$$\text{RELERR} = 0 \ \textit{and} \ \text{ABSERR} = 0$$

are considered unforgivable errors and the code returns with IFLAG = 3. A normal return has IFLAG = 1 and the apparently successful computation of the solution at A + DA. To control the amount of work, an internal function counter permits only 3000 function evaluations per call, and an excess causes a return with IFLAG = 2. Nothing is wasted because of the way the routine returns. To continue, just call the routine again. If the code thinks it needs a step size too small for the working precision, it returns with IFLAG = 4.

The quantities A, Y, and H are all used for output as well as input, and hence must be variables in the calling program.

The code is organized to facilitate the production of a table of solution values. The following example shows a typical use of RKF. Suppose we want to integrate

$$Y_1' = Y_1,$$
$$Y_2' = -Y_2,$$
$$Y_1(0) = 1,$$
$$Y_2(0) = 1$$

from $x = 0$ to $x = 1$, printing the results at $x = 1$. The solutions are obviously $Y_1(x) = e^x$, $Y_2(x) = e^{-x}$. Let us take RELERR = 0, ABSERR = 1.E−6, and HMX = 1.

```
      EXTERNAL F
      DIMENSION Y(2)
      Y(1) = 1.0
      Y(2) = 1.0
      A = 0.0
      H = 0.2
      HMX = 1.0
      CALL RKF (A, Y, 2, F, 1.0, H, HMX, 1.E-6, 0.0, IFLAG)
      GO TO (1, 2, 2, 2), IFLAG
    1 PRINT 100, A, Y(1), Y(2), H
  100 FORMAT (F9.6, 1P3E15.7)
      STOP
```

```
  2 PRINT 101, A, IFLAG
101 FORMAT (F9.6, I15)
    STOP
    END
    SUBROUTINE F(X, Y, YP)
    DIMENSION Y(1), YP(1)
    YP(1) = Y(1)
    YP(2) = -Y(2)
    RETURN
    END
```

This code yields

1.000000	2.718270E+00	3.678794E−01	1.130731E−01.

If we wanted to print the solution at $x = 1, 2, 3, \ldots, 10$, all we would need to do is insert the statement

```
DO  1  I = 1, 10
```

just before the CALL RKF statement. Notice that the optimal H is used on all calls after the first. The results are

1.000000	2.718270E+00	3.678794E−01	1.130731E−01
2.000000	7.389053E+00	1.353344E−01	8.441198E−02
3.000000	2.008560E+01	4.978636E−02	6.327099E−02
3.674215	4		

Here the code did not reach the desired value of $x = 10$. This was signaled by an IFLAG of 4. The last successful point at which the solution was computed was $x = 3.674215$

```
      SUBROUTINE RKF(A,Y,N,F,DA,H,HMX,ABSERR,RELERR,IFLAG)
C
C  THE CODE INTEGRATES A SYSTEM OF FIRST ORDER ORDINARY DIFFERENTIAL
C  EQUATIONS BY RUNGE-KUTTA-FEHLBERG METHOD WITH AUTOMATIC ESTIMATION
C  OF LOCAL ERROR AND STEP SIZE ADJUSTMENT.
C
C         A - ON INPUT THE INITIAL VALUE OF THE INDEPENDENT VARIABLE.
C             ON OUTPUT THE LAST VALUE AT WHICH A SOLUTION WAS
C             SUCCESSFULLY COMPUTED.  NORMAL OUTPUT HAS INPUT VALUE
C             OF A INCREASED BY DA.
C         Y - ON INPUT THE VECTOR OF INITIAL VALUES OF THE DEPENDENT
C             VARIABLES.  ON OUTPUT THE VECTOR OF COMPUTED SOLUTIONS
C             AT OUTPUT VALUE OF A.
C         N - NUMBER OF EQUATIONS TO BE INTEGRATED.
C         F - SUBROUTINE OF THE FORM F(A,Y,YP) WHICH ACCEPTS INDEPEN-
C             DENT VARIABLE A AND A VECTOR OF DEPENDENT VARIABLES
C             Y AND FURNISHES A VECTOR OF VALUES OF THE DERIVATIVES
C             YP.  USE DIMENSIONS Y(1), YP(1).  IN VECTOR FORM THE
```

```
C             DIFFERENTIAL EQUATIONS SOLVED ARE
C
C                   DY/DX=YP, Y(A)=Y.
C
C        DA - INTEGRATION IS TO PROCEED FROM A TO  A + DA.  DA CAN BE
C             NEGATIVE.
C        H - NOMINAL STEP SIZE ON INPUT.  ON OUTPUT IT IS REPLACED BY
C             OPTIMAL VALUE CURRENTLY BEING USED BY THE CODE.  H CAN
C             BE NEGATIVE.
C        HMX - UPPER BOUND ON STEP SIZE TO BE USED, MUST BE POSITIVE.
C        ABSERR,RELERR - BOUNDS ON LOCAL ERROR PERMITTED, RELATIVE TO A
C             UNIT CHANGE IN THE INDEPENDENT VARIABLE.  EACH COMPONENT
C             OF THE COMPUTED SOLUTION Y(I) OBTAINED IN A STEP OF LENGTH
C             ABS(H) MUST PASS THE TEST
C
C                   ABS(ESTIMATED LOCAL ERROR).LE.
C                   ABS(H)*(RELERR*ABS(Y(I))+ABSERR)
C
C   A FLAG, IFLAG, IS PROVIDED AS AN OUTPUT QUANTITY.  IT MAY ASSUME
C   THE VALUES 1-4 WHERE
C
C        IFLAG = 1  FOR NORMAL RETURN--REACHED  A+DA.
C
C              = 2  IF MORE THAN 3000 FUNCTION EVALUATIONS ARE NEEDED.
C
C              = 3  FOR ILLEGAL INPUT VALUES.
C
C              = 4  IF CODE NEEDS TO TAKE TOO SMALL A STEP FOR
C                   COMPUTER WORD LENGTH.
C
C   A, Y, AND H MUST BE VARIABLES IN THE CALLING PROGRAM SINCE THEY
C   ARE ALSO USED FOR OUTPUT.  Y MUST BE DIMENSIONED IN THE CALLING
C   PROGRAM.
C
      DIMENSION Y(N)
C
C   THE CODE IS SET UP TO SOLVE SYSTEMS OF UP TO 10 DIFFERENTIAL
C   EQUATIONS.  IF MORE EQUATIONS ARE DESIRED, ONLY THE NEXT TWO
C   STATEMENTS NEED BE CHANGED.
C
      DIMENSION YTEMP(10),TEMP(10),R(10)
      REAL K1(10),K2(10),K3(10),K4(10),K5(10),K6(10)
C
C   SET U TO APPROXIMATELY THE UNIT ROUND-OFF OF SPECIFIC MACHINE.
C   (HERE IBM 360/67)
C
      U=9.0E-7
C
C   TEST INPUT DATA AND INITIALIZE.
C
      IF(RELERR.LT.0.0.OR.ABSERR.LT.0.0)GO TO 18
      IF(RELERR+ABSERR.EQ.0.0)GO TO 18
      IF(HMX.LE.0.0)GO TO 18
      B=A+DA
      IF(ABS(DA).LE.13.0*U*AMAX1(ABS(A),ABS(B)))GO TO 19
      HMAX=AMIN1(HMX,ABS(DA))
      IF(ABS(H).LE.13.0*U*ABS(A))H=HMAX
      KOUNT=0
      IADJUS=0
C
C   LIMIT H TO HMAX AND ADJUST STEP SIZE TO MAKE OUTPUT POINT IF
C   APPROPRIATE.
C
    3 H=SIGN(AMIN1(ABS(H),HMAX),DA)
      IF(ABS(B-A).GT.1.25*ABS(H))GO TO 4
      HKEEP=H
C
C   CODE REALIZES IT HAS REACHED OUTPUT POINT WHEN IT TAKES A
C   SUCCESSFUL STEP WITH VARIABLE IADJUS=1.
```

```
C
      IADJUS=1
      H=B-A
C
C  BEGIN COMPUTATION OF STEP.
C
    4 CALL F(A,Y,K1)
      KOUNT=KOUNT+1
    5 CONTINUE
      DO 6 I=1,N
    6 YTEMP(I)=Y(I)+0.25*H*K1(I)
      ARG=A+0.25*H
      CALL F(ARG,YTEMP,K2)
      DO 7 I=1,N
    7 YTEMP(I)=Y(I)+H*(K1(I)*(3.0/32.0)+K2(I)*(9.0/32.0))
      ARG=A+H*(3.0/8.0)
      CALL F(ARG,YTEMP,K3)
      DO 8 I=1,N
    8 YTEMP(I)=Y(I)+H*(K1(I)*(1932.0/2197.0)-K2(I)*(7200.0/2197.0)
     1 +K3(I)*(7296.0/2197.0))
      ARG=A+H*(12.0/13.0)
      CALL F(ARG,YTEMP,K4)
      DO 9 I=1,N
    9 YTEMP(I)=Y(I)+H*(K1(I)*(439.0/216.0)-8.0*K2(I)
     1 +K3(I)*(3680.0/513.0)-K4(I)*(845.0/4104.0))
      ARG=A+H
      CALL F(ARG,YTEMP,K5)
      DO 10 I=1,N
   10 YTEMP(I)=Y(I)+H*(-K1(I)*(8.0/27.0)+2.0*K2(I)-K3(I)*
     1 (3544.0/2565.0)+K4(I)*(1859.0/4104.0)-K5(I)*(11.0/40.0))
      ARG=A+0.5*H
      CALL F(ARG,YTEMP,K6)
      DO 11 I=1,N
      TEMP(I)=K1(I)*(25.0/216.0)+K3(I)*(1408.0/2565.0)+K4(I)*
     1 (2197.0/4104.0)-0.2*K5(I)
   11 YTEMP(I)=Y(I)+H*TEMP(I)
C
C  NOW YTEMP(I) IS TENTATIVE RESULT OF THE STEP.  COMPUTE R(I),
C  THE ESTIMATED LOCAL ERROR RELATIVE TO A UNIT CHANGE IN THE
C  INDEPENDENT VARIABLE.
C
      DO 12 I=1,N
   12 R(I)=K1(I)/360.0-K3(I)*(128.0/4275.0)-K4(I)*(2197.0/75240.0)+
     1 K5(I)/50.0+K6(I)*(2.0/55.0)
C
C  TEST FOR ACCURACY.
C
      RATIO=0.0
      DO 13 I=1,N
      TR=ABS(R(I))/(RELERR*ABS(YTEMP(I))+ABSERR)
   13 RATIO=AMAX1(RATIO,TR)
C
C  RATIO.GT.1.0 MEANS REJECT STEP AND REPEAT.  RATIO IS USED IN EITHER
C  EVENT FOR STEP ADJUSTMENT.
C
      IF(RATIO.GT.1.0)GO TO 15
C
C  ACCEPT RESULT OF STEP.
C
      DO 14 I=1,N
   14 Y(I)=YTEMP(I)
      A=A+H
C
C  IF WE HAVE REACHED A+DA, RETURN.
C
      IF(IADJUS.EQ.1)GO TO 16
C
C  REFINE OR CHOOSE THE NEXT STEP.  RATIO IS ALTERED IF NECESSARY
```

```
C   SO THAT THE INCREASE IS LIMITED TO A FACTOR OF 5 AND THE DECREASE
C   TO A FACTOR OF 1/10.
C
      RATIO=AMAX1(RATIO,6.5536E-4)
 15   RATIO=AMIN1(RATIO,4096.0)
      H=0.8*H/SQRT(SQRT(RATIO))
      IF(ABS(H).LE.13.0*U*ABS(A))GO TO 19
      KOUNT=KOUNT+5
      IF(KOUNT.GE.2995)GO TO 17
C
C   IF RATIO.LE.1.0, THE STEP WAS SUCCESSFUL SO WE START ANOTHER
C   STEP.  OTHERWISE WE ARE REPEATING WITH A SMALLER H.
C
      IF(RATIO.LE.1.0)GO TO 3
      IADJUS=0
      GO TO 5
 16   IFLAG=1
      H=HKEEP
      RETURN
 17   IFLAG=2
      RETURN
 18   IFLAG=3
      RETURN
 19   IFLAG=4
      RETURN
      END
```

SYSTEMS OF LINEAR EQUATIONS

The code LASSOL solves a system of linear equations using the method of Gaussian elimination with partial pivoting and row equilibration as described in III. To make the code easier to follow, the scaling for row equilibration has been done explicitly and rows have been physically interchanged in the pivoting. To use the code, the system must be written in the form

$$\begin{array}{c} a_{11}x_1 + a_{12}x_2 + \cdots + a_{1N}x_N = b_1 \\ \vdots \\ a_{N1}x_1 + a_{N2}x_2 + \cdots + a_{NN}x_N = b_N \end{array}$$

or using matrix notation,

$$Ax = b$$

where

$$A = \begin{bmatrix} a_{11} & a_{12} & \cdots & a_{1N} \\ a_{21} & a_{22} & \cdots & a_{2N} \\ \cdot & & & \cdot \\ \cdot & & & \cdot \\ \cdot & & & \cdot \\ a_{N1} & a_{N2} & \cdots & a_{NN} \end{bmatrix}, \quad x = \begin{bmatrix} x_1 \\ x_2 \\ \cdot \\ \cdot \\ \cdot \\ x_N \end{bmatrix}, \quad b = \begin{bmatrix} b_1 \\ b_2 \\ \cdot \\ \cdot \\ \cdot \\ b_N \end{bmatrix}.$$

The call list is

LASSOL(N, A, B, M, X, R, IFLAG)

The parameters of this list used for input are:

N—the number of equations and unknowns
A—the N by N matrix of coefficients
B—the vector right hand side
M—dimension of A in the calling program

Neither A nor B is altered by the program. There are three outputs from LASSOL. The first is the solution vector X and the second is the maximum residual R. The last quantity is a flag, IFLAG, provided to warn the user that he has a problem which is "apparently" singular. By this we mean that the problem on which the machine is working (probably different from the problem we originally had in mind because of roundoff errors) is, to within the limits imposed by the computer arithmetic and this algorithm, singular. For a non-singular problem, IFLAG = 1; for a singular one, IFLAG is set to 2 and control is returned to the calling program. The arrays A, B, and X must be dimensioned in the calling program. It is sometimes convenient to have the matrix A stored in an array larger than necessary, so we allow the matrix A to be stored in an $M \times M$ array where, of course, $M \geq N$.

The residual vector is calculated in double precision. This is necessary to insure a meaningful residual; if one wishes to write the code wholly in single precision, he should delete the residual computation entirely. The code uses a work array AB into which A and B are copied. If one deletes the residual computation and he does not mind destroying A and B, he can rewrite the code so as to do without AB and so cut his storage in half.

As an example let us solve the system

$$2.0x_1 + 3.0x_2 = 7.0,$$
$$3.0x_1 + 4.0x_2 = 8.0.$$

In the notation above,

$$A = \begin{bmatrix} 2.0 & 3.0 \\ 3.0 & 4.0 \end{bmatrix}, \qquad B = \begin{bmatrix} 7.0 \\ 8.0 \end{bmatrix}$$

and N = 2. We shall store A in a 3×3 array for illustrative purposes. The following program used LASSOL and obtained the results indicated:

```
      DIMENSION A(3, 3), B(2), X(2)
      A(1, 1) = 2.0
      A(2, 1) = 3.0
      A(1, 2) = 3.0
      A(2, 2) = 4.0
      B(1) = 7.0
      B(2) = 8.0
      N = 2
      M = 3
      CALL LASSOL (N, A, B, M, X, R, IFLAG)
      PRINT 1, IFLAG, (X(I), I = 1, N), R
    1 FORMAT (I5, 1P3E17.7)
      STOP
      END

    1  -0.3999997E+01  0.4999998E+01  0.9536743E-6
```

In words, the solution to the set of equations is $x_1 = -3.999997$ and $x_2 = 4.999998$, with the maximum residual $R = 9.5 \times 10^{-7}$. IFLAG = 1, indicating that the matrix A is apparently nonsingular.

```
      SUBROUTINE LASSOL(N,A,B,M,X,R,IFLAG)
C
C  LASSOL SOLVES A SYSTEM OF N LINEAR EQUATIONS IN N UNKNOWNS,
C  AX=B, USING GAUSSIAN ELIMINATION WITH PARTIAL PIVOTING AND
C  ROW EQUILIBRATION.  INPUT QUANTITIES ARE
C
C         N - NUMBER OF EQUATIONS AND UNKNOWNS.
C         A - N BY N MATRIX OF COEFFICIENTS STORED IN AN M BY M
C             ARRAY.
C         B - VECTOR OF RIGHT HAND SIDES.
C         M - DIMENSION OF A IN THE CALLING PROGRAM.
C
C  OUTPUT QUANTITIES ARE
C
C         X - SOLUTION VECTOR.
C         R - LARGEST RESIDUAL IN MAGNITUDE.
C         IFLAG = 1 FOR NORMAL RETURN.
C               = 2 IF MATRIX APPEARS SINGULAR TO THE CODE.  CONTROL
C                   IS RETURNED TO THE CALLING PROGRAM.
C
C  THE ARRAYS A, B, AND X MUST BE DIMENSIONED IN THE
C  CALLING PROGRAM.
C
      DIMENSION A(M,M),B(N),X(N),
C
C  LASSOL IS WRITTEN TO SOLVE SYSTEMS OF SIZE UP TO 10 BY 10.
C  DIMENSION OF WORKING STORAGE AB MUST BE GREATER THAN OR
C  EQUAL TO (N,N+1).  TO CHANGE THE SIZE OF PROBLEMS THAT CAN
C  BE HANDLED, ONLY THE NEXT LINE NEED BE MODIFIED.
C
     1  AB(10,11)
      DOUBLE PRECISION SUMR
      NP1=N+1
      NM1=N-1
C
C  FORM THE N BY (N+1) MATRIX AB, THE FIRST N COLUMNS OF WHICH
```

```
C   ARE A AND THE REMAINING COLUMN B.  CALCULATE SCALE FACTORS
C   AND SCALE AB.
C
      DO 3 I=1,N
      ROWMAX=0.0
      DO 1 J=1,N
 1    ROWMAX=AMAX1(ROWMAX,ABS(A(I,J)))
      IF(ROWMAX.EQ.0.0)GO TO 14
      SCALE=1.0/ROWMAX
      DO 2 J=1,N
 2    AB(I,J)=A(I,J)*SCALE
 3    AB(I,NP1)=B(I)*SCALE
C
C  BEGIN BASIC ELIMINATION LOOP.  ROWS OF AB ARE PHYSICALLY
C  INTERCHANGED IN ORDER TO BRING ELEMENT OF LARGEST MAG-
C  NITUDE INTO PIVOTAL POSITION.
C
      DO 9 K=1,NM1
      BIG=0.0
      DO 4 I=K,N
      TEMPB=ABS(AB(I,K))
      IF(BIG.GE.TEMPB)GO TO 4
      BIG=TEMPB
      IDXPIV=I
 4    CONTINUE
      IF(BIG.EQ.0.0)GO TO 14
      IF(IDXPIV.EQ.K)GO TO 6
C
C  PIVOT IS IN ROW IDXPIV.  INTERCHANGE ROW IDXPIV WITH ROW K.
C
      DO 5 I=K,NP1
      TEMPI=AB(K,I)
      AB(K,I)=AB(IDXPIV,I)
 5    AB(IDXPIV,I)=TEMPI
 6    KP1=K+1
C
C  ELIMINATE X(K) FROM EQUATIONS K+1,K+2,...,K+N.
C
      DO 8 I=KP1,N
      QUOT=AB(I,K)/AB(K,K)
      DO 7 J=KP1,NP1
 7    AB(I,J)=AB(I,J)-QUOT*AB(K,J)
 8    CONTINUE
 9    CONTINUE
C
C  BEGIN CALCULATION OF SOLUTION X USING BACK SUBSTITUTION.
C
      IF(AB(N,N).EQ.0.0)GO TO 14
      X(N)=AB(N,NP1)/AB(N,N)
      DO 11 IB=2,N
      I=NP1-IB
      IP1=I+1
      SUM=0.0
      DO 10 J=IP1,N
10    SUM=SUM+AB(I,J)*X(J)
11    X(I)=(AB(I,NP1)-SUM)/AB(I,I)
C
C  CALCULATE MAXIMUM RESIDUAL R.
C
      R=0.0
      DO 13 I=1,N
      SUMR=0.0D0
      DO 12 J=1,N
12    SUMR=SUMR+DBLE(A(I,J))*DBLE(X(J))
13    R=AMAX1(R,ABS(SNGL(SUMR-DBLE(B(I)))))
      IFLAG=1
      RETURN
14    IFLAG=2
      RETURN
      END
```

REFERENCES

With a few exceptions, we list those books and papers referred to in the text. The exceptions are of less general interest and are referenced only once, so they are given at the point they arise. The reader may also be interested in sources of algorithms. The principal published sources are the journals: Communications of the Association for Computing Machinery (CACM), the British Computer Journal, the Scandinavian journal BIT, and Numerische Mathematik. The CACM has a convenient index by subject to algorithms for the period 1960–1968 in vol. 11, 1968, pp. 827–830, which includes algorithms in other journals, too. Algorithms are being prepublished in Numerische Mathematik before their publication in book form as volumes in a handbook of automatic computation. At this writing, one volume has appeared on linear algebra; see Wilkinson and Reinsch [48]. Some books are beginning to include algorithms, e.g., [3, 13, 14, 19, 20, 22, 40, 43, 48].

1. Abramowitz, M., and I. A. Stegun, eds., *Handbook of Mathematical Functions*, No. 55 in NBS Applied Math. Series, U.S. Govt. Printing Off., Wash., D.C., 1966.
2. Ahlberg, J. H., E. N. Nilson, and J. L. Walsh, *The Theory of Splines and Their Applications*, Academic Press, New York, 1967.
3. Bareiss, E. H., The numerical solution of polynomial equations and the resultant procedures, Chap. 10 in *Mathematical Methods for Digital Computers*, Vol. 2, A. Ralston and H. S. Wilf, eds., John Wiley & Sons, New York, 1967.
4. Bateman, H., *Partial Differential Equations of Mathematical Physics*, Cambridge Univ. Press, London, 1964.
5. Bers, L., *Calculus*, Holt, Rinehart, and Winston, New York, 1969.
6. Björk, A., Solving least squares problems by Gram-Schmidt orthogonalization, BIT 7 (1967) 1–21.
7. Businger, P., and G. Golub, Linear least squares solutions by Householder transformations, Num. Math. 7 (1965) 269–276.
8. Casaletto, J., M. Pickett, and J. R. Rice, A comparison of some numerical integration programs, Purdue Univ. Computer Sciences Dept. report CSD TR 37.
9. Ceschino, F., and J. Kuntzmann, *Numerical Solution of Initial Value Problems*, Prentice-Hall, Englewood Cliffs, N.J., 1966.
10. Chandrasekhar, S., *Radiative Transfer*, Dover, New York, 1960.
11. Coddington, E. A., *An Introduction to Ordinary Differential Equations*, Prentice-Hall, Englewood Cliffs, N.J., 1961.
12. Davis, P. J., Orthonormalizing codes in numerical analysis, Chap. 10 in *Survey of Numerical Analysis*, J. Todd, ed., McGraw-Hill, New York, 1962.
13. Davis, P. J., and P. Rabinowitz, *Numerical Integration*, Blaisdell, Waltham, Mass., 1967.
14. Dekker, T. J., Finding a zero by means of successive linear interpolation, in *Constructive Aspects of the Fundamental Theorem of Algebra*, B. Dejon and P. Henrici, eds., Wiley-Interscience, London, 1969.
15. Fehlberg, E., Klassiche Runge-Kutta-Formeln vierter und niedrigerer Ordnung

mit Schrittweiten-Kontrolle und ihre Anwendung auf Wärmeleitungsprobleme, Computing *6* (1970) 61–71.

16. Fike, C. T., *Computer Evaluation of Mathematical Functions*, Prentice-Hall, Englewood Cliffs, N.J., 1968.
17. Forsythe, G. E., Generation and use of orthogonal polynomials for data-fitting with a digital computer, J. SIAM *5* (1956) 74–88.
18. Forsythe, G. E., What is a satisfactory quadratic equation solver, in *Constructive Aspects of the Fundamental Theorem of Algebra*, B. Dejon and P. Henrici, eds., Wiley-Interscience, London, 1969.
19. Forsythe, G. E., and C. B. Moler, *Computer Solution of Linear Algebraic Systems*, Prentice-Hall, Englewood Cliffs, N.J., 1967.
20. Fox, P. A., DESUB: Integration of a first-order system of ordinary differential equations, Chap. 9 in *Mathematical Software*, J. R. Rice, ed., Academic Press, New York, 1971.
21. Fröberg, C.-E., *Introduction to Numerical Analysis*, Addison-Wesley, Reading, Mass., 1969.
22. Gear, C. W., *Numerical Initial Value Problems in Ordinary Differential Equations*, Prentice-Hall, Englewood Cliffs, N.J., 1971.
23. Henrici, P., *Elements of Numerical Analysis*, John Wiley & Sons, New York, 1964.
24. Householder, A. S., *The Numerical Treatment of a Single Nonlinear Equation*, McGraw-Hill, New York, 1970.
25. Hull, T. E., W. H. Enright, B. M. Fellen, and A. E. Sedgwick, Comparing numerical methods for ordinary differential equations, SIAM J. on Numer. Anal. *9* (1972) 603-637.
26. Isaacson, E., and H. B. Keller, *Analysis of Numerical Methods*, John Wiley & Sons, New York, 1966.
27. Kahaner, D. K., Comparison of numerical quadrature formulas, Los Alamos Scientific Laboratory report LA-4137.
28. Krogh, F. T., Efficient algorithms for polynomial interpolation and numerical differentiation, Math. Comp. *24* (1970) 185–190.
29. Leighton, R. B., *Principles of Modern Physics*, McGraw-Hill, New York, 1959.
30. Lyness, J. N., Notes on the adaptive Simpson quadrature routine, JACM *16* (1969) 483–495.
31. McKeeman, W. M., Algorithm 145, Adaptive numerical integration by Simpson's rule, CACM *5* (1962) 604.
32. McKeeman, W. M., and L. Tesler, Algorithm 182, Nonrecursive adaptive integration, CACM *6* (1963) 315.
33. Moon, P., and D. E. Spencer, *Field Theory for Engineers*, Van Nostrand, Princeton, N.J., 1961.
34. Noble, B., *Applications of Undergraduate Mathematics in Engineering*, Macmillan, New York, 1967.
35. Noble, B., *Applied Linear Algebra*, Prentice-Hall, Englewood Cliffs, N.J., 1969.
36. Ortega, J. M., and W. C. Rheinboldt, *Iterative Solution of Nonlinear Equations in Several Variables*, Academic Press, New York, 1970.
37. Ostrowski, A. M., *Solution of Equations and Systems of Equations*, Academic Press, New York, 1966.
38. Purcell, E. J., *Calculus with Analytic Geometry*, Appleton-Century-Crofts, New York, 1972.
39. Rice, J. R., Experiments on Gram-Schmidt orthogonalization, Math. Comp. *20* (1966) 325–328.
40. Stroud, A. H., and D. Secrest, *Gaussian Quadrature Formulas*, Prentice-Hall, Englewood Cliffs, N.J., 1966.
41. Thomas, G. B., *Calculus and Analytic Geometry*, Addison-Wesley, Reading, Mass., 1968.

42. Traub, J. F., *Iterative Methods for the Solution of Equations*, Prentice-Hall, Englewood Cliffs, N.J., 1964.
43. Traub, J. F., The solution of transcendental equations, Chap. 9 in *Mathematical Methods for Digital Computers*, Vol. 2, A. Ralston and H. S. Wilf, eds., John Wiley & Sons, New York, 1967.
44. Uspensky, J. V., *Theory of Equations*, McGraw-Hill, New York, 1948.
45. Wilkinson, J. H., *Rounding Errors in Algebraic Processes*, Prentice-Hall, Englewood Cliffs, N.J., 1963.
46. Wilkinson, J. H., *The Algebraic Eigenvalue Problem*, Oxford Univ. Press, London, 1965.
47. Wilkinson, J. H., The solution of ill-conditioned linear equations, chap. 3 in *Mathematical Methods for Digital Computers*, Vol. 2, A. Ralston and H. S. Wilf, eds., John Wiley & Sons, New York, 1967.
48. Wilkinson, J. H., and C. Reinsch, *Handbook for Automatic Computation*, Vol. II *Linear Algebra*, Springer, New York, 1971.